Electrical Inspection Workbook
By Tom Henry

While every precaution has been taken in the preparation of this material to ensure accuracy and reliability of the information, instructions, and directions, it is in no way to be construed as a guarantee. The author and publisher assumes no responsibility for errors or omissions. Neither is any liability assumed from the use of the information contained herein in case of misinterpretations, human error or typographical errors.

National Electrical Code® and NEC® are Registered Trademarks of the National Fire Protection Association, Inc., Quincy, MA.

Second printing December 1999

HENRY PUBLICATIONS SINCE 1985

ISBN 0 - 945495 - 58 - 7

CONTENT

Preface

With my experience as an electrical expert on personal injuries with a law firm and having worked as an electrical inspector, I truly see the need for a workbook such as this.

The big problem today that I see in the industry is, in some cases we have complex decisions being made by people with little or no training and possibly no understanding of the implications of their decisions.

The complexity of the technology and the codes changing rapidly requires a higher skill level.

The inspector should be one of the most highly skilled professionals in the construction industry. Often he must possess the knowledge of an engineer, educator, business consultant, etc.

At a time when the country is experiencing its greatest need for highly competent inspectors, local government budget cuts are pushing some inspectors to becoming "the jack-of-all-trades, master of none."

Inspection requirements vary in different areas of the country. Some areas require an inspection of the water system before you can take occupancy. Other areas require a termite inspection before the sale of the property. While other areas may require an inspection of the foundation or roof, etc. The *most important* inspection for safety, an *electrical* inspection, is not even required in some areas of the country! Have you ever read in a newspaper where water started a fire? Has your local newspaper ever contained an article where a termite caused an electrocution? Electricity has been known to start fires and electrocute people.

Our society is becoming more litigious, and the decisions the inspector makes will be examined more closely. Municipalities that cut inspection corners will pay the price down the road, either in more fires or in lawsuits. Insurance companies have made several recent attempts to sue, citing failure to *properly* inspect buildings for hazards.

Inspection is a very demanding career, and the technical expertise required to handle the job means that individuals today tend to be more educated than before. Education is the key.

To survive in an era when voters demand more government services for less money, today inspectors must be much more than code enforcers. They must prove their value to the community. They must be seen as partners and consultants in fire and accident prevention. The inspector's job is to keep the community's assets intact, to ensure continuity of operations.

The inspector is a consultant who's already been paid for. Would you hire yourself, and would you get your money's worth? If not, you can expect the community to turn elsewhere.

ACKNOWLEDGEMENTS

Special appreciation is expressed to the following individuals for reviewing this workbook. Their analysis and recommendations for changes and additions have helped the author reach the initial goal of a quality book on electrical inspections.

Paul D. Ward

Paul is currently the Chief Building Official for the City of Southlake, Texas. Responsibilities include the enforcement, interpretation and regulation of building, plumbing, electrical, mechanical, zoning, and sign codes.

Presently Paul serves as secretary for the Building Board of Appeals. This board makes decisions concerning new building products and substandard or dangerous buildings within the city's jurisdiction.

Paul was employed by the City of Abilene, Texas as an electrical inspector from 1989 to 1992.

- Certified Building Official CABO
- Block Master Electrical License
- Certified Electrical Inspector ICBO
- Certified Electrical Inspector IAEI
- Certified Plumbing Inspector
- Certified Building Inspector ICBO
- Certified Mechanical Inspector ICBO
- Certified Combination Inspector ICBO

Clarence N. Wilson

Presently Clarence is the Chief Electrical Inspector for the City of Orlando, Florida. Responsibilities include supervision of electrical inspectors, review of electrical plans for Code conformance, review sign plans for permitting, review permit applications, review requirements for electrical examination license, etc.

Clarence has served as chairman of the Central Florida IAEI chapter and at the present time is a state board member.

Very active in community affairs motivating young people towards education.

- Board member Electrical Council of Florida
- School Advisory Council Chairman Mid Florida Tech
- Electrical Advisory Board Chairman Mid Florida Tech
- Florida Electrical Apprentice Training Instructor
- Certified Building Official BOAF
- Certified Chief Electrical Inspector SBCCI
- Certified Plans Examiner SBCCI
- Certified Vocational Instructor State of Florida
- Block Master Electrical License
- Member National Fire Protection Association

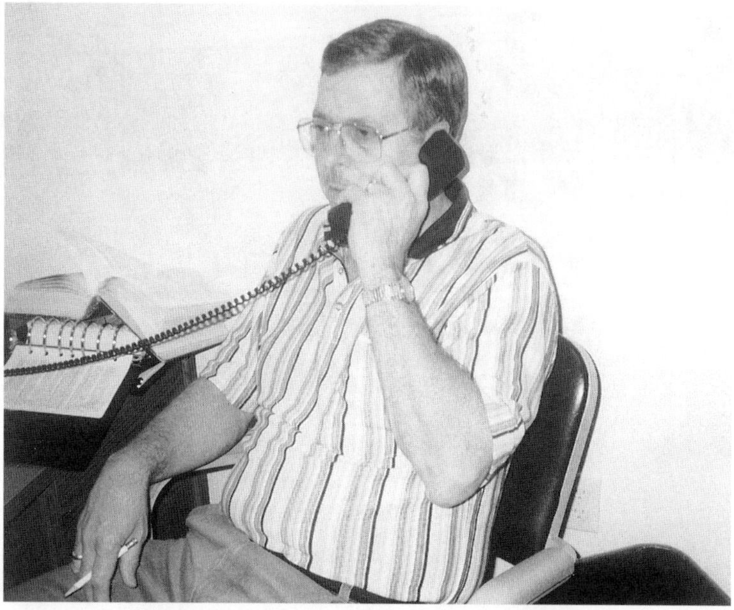

James E. Lewis

At the present Jim teaches classes on the National Electrical Code and Business Law for the electrical contractor throughout the United States.

Jim also serves as the Training Director for *Tom Henry's* "Learn to be an Electrician" program which has enrollment from throughout the world.

From 1980 to 1986 he was the Chief Electrical Inspector for Walt Disney World and EPCOT Center. Responsibilities included new construction, additions and rehabs of existing property, safety inspections on emergency systems, plans examination, budget preparation, and supervision.

Jim was instrumental in the forming of the IAEI Central Florida Chapter and served as the first secretary.

•Certified Chief Electrical Inspector SBCCI
•Certified Plans Examiner SBCCI
•Certified Commercial Electrical Inspector SBCCI
•Certified Vocational Instructor State of Florida
•Registered Electrical Contractor State of Florida
•Master Electrical License State of Alabama
•Master Electrical License State of Georgia
•Member National Fire Protection Association
•Member International Association of Electrical Inspectors
•Member of the Electrical Council of Florida

CHAPTER 1

THE INSPECTOR

90-4

THE AUTHORITY HAVING JURISDICTION

The authority having jurisdiction is inspectors employed by a government agency (federal, state, county, city) which has authority to enforce local ordinances and code regulations concerning product installations. The authority having jurisdiction has the responsibility for (1) making interpretations of the rules, (2) deciding upon the approval of equipment and materials, and (3) granting the special permission contemplated in a number of the rules.

The final responsibility for interpreting the rules and requirements given in the Code is the "authority having jurisdiction", and that "authority" is generally the local electrical inspector.

Where the National Electrical Code is adopted by law, ordinance, or regulation as referenced to electrical installations and inspections, it has the force of law. It must be realized, however, that when any code or standard is adopted in any city, town, county, or other political subdivision certain specific **ADDITIONS OR DELETIONS** may be made a part of the adopted ordinance.

Inspectors are required in some cases to make rulings on local conditions such as soil or weather. The sealing of conduits for a gas pump in Alaska may require double sealing for the extremely cold atmospheres where seals may fail. In Florida it's better to use a PVC conduit below grade due to the soil conditions. Some local Codes require this. This decision would have to be made by the inspector in that area.

The inspector is required to evaluate all the factors in each case, and give the local interpretation, or waive specific Code requirements where **EQUIVALENT SAFETY CAN BE ACHIEVED BY OTHER MEANS.**

Code section 90-4 does give the inspector some latitude in making decisions.

Most inspection authorities do not like the extra pressure that 90-4 would place upon them *if they made special exceptions*, since unforeseen resulting hazards could bring legal action against them.

Inspectors are NOT generally permitted to make whimsical rulings. Whimsical rulings by inspectors could not be lawfully enforced, but ordinances can be.

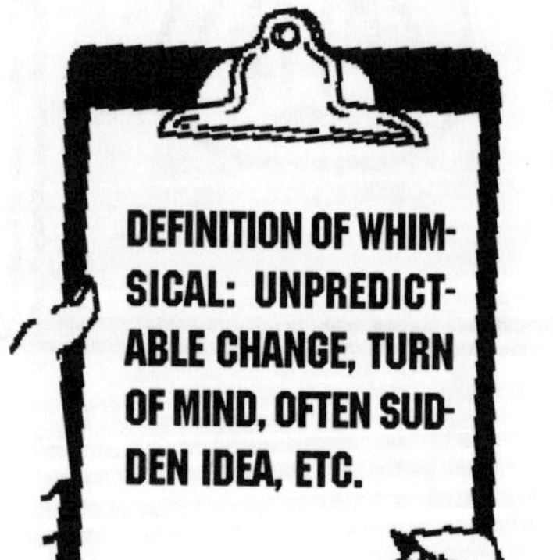

DEFINITIONS are very important to the inspector.

The definition of **APPROVED** is acceptable to the authority having jurisdiction.

The word **APPROVED** rather than *listed* allows the inspector to impose even stricter requirements where warranted.

If approved wiring methods are used to install the electrical wiring in a building, but low quality materials are used, the complete installation may still be dangerous.

IDENTIFIED means generally recognizable as suitable for the application. It does not mean listed, but it could be. All products listed for a particular application are necessarily identified for that use but not all products identified for a particular use have been specifically listed.

LISTED equipment or materials included in a list published by an organization acceptable to the inspector and concerned with product evaluation, that maintains periodic inspection of production of listed equipment or materials, and whose listing states either that the equipment or material meets appropriate designed standards or has been tested and found suitable for use in a specific manner.

The means for identifying listed equipment may vary for each organization concerned with product evaluation, some of which do not recognize equipment as listed unless it is also **LABELED**. The inspector should utilize the system employed by the listing organization to identify a listed product.

Underwriter's Laboratories (UL) is a nonprofit organization supported by the manufacturers who submit merchandise; that is, for testing the merchandise, Underwriters' Laboratories charges a fee, which pays for the inspectors' expenses and supports the laboratories in general.

A UL label assures the user that the manufacturer of the item has submitted to the laboratories for testing samples that were found to meet the required minimum safety standards. It is then presumed that the manufacturer will maintain the same quality in future production of the same item.

An item approved by UL, however does not mean that the item is approved for all uses. Rather, a UL label means that the item as labelled is safe only for the purpose for which it was intended. The UL label means the item meets the minimum safety requirements; one kind may meet the requirements, while another may far exceed them. Therefore, quality cannot really be judged from the UL label alone.

The UL **GREEN BOOK** "Electrical Construction Directory" is a must for the electrical inspector.

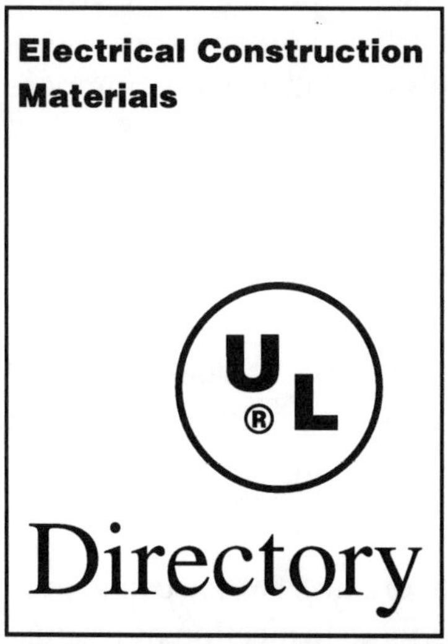

The directory may be used for several purposes, some of which are: (a) To obtain the names of companies which are able to provide products bearing a Listing Mark or Classification Marking. (b) To obtain information pertaining to the form and nature of the Listing Mark or Classification Marking to be used for a specific class or category of product. (c) To obtain information relating to limitations or special conditions applying to the product or (d) To obtain the titles and designations of Standards that have been used for investigation of products in a specific category. The companies whose names appear in the Directory are qualified to provide products that bear the **UL** Mark but may not be the actual manufacturer.

The UL Green Book states PVC Schedule 40 must NOT be used where subject to physical damage. The Green Book states wireways, auxiliary gutters, and associated fittings are suitable for use as equipment grounding conductors.

A UL listed pressure terminal connector for more than one conductor will be marked either on the shipping carton or on the connector itself.

If the products pass the exhaustive tests in accordance with established standards, they are listed in the UL official published list. Such UL approved items usually have a UL **LABEL** attached directly to the product. In some cases, this label is molded or stamped into the item, into bakelite, steel, or porcelain parts of wiring devices.

The electrical inspector is required to be very versatile in dealing with the sophisticated electrical equipment and systems found in various segments of the electrical industry which are in residential, commercial, and industrial applications.

A successful inspection program entails the embracing of the following items as law:

1. Licensing
2. Electrical permits
3. Electrical codes
4. Electrical standards
5. Enforcement
6. Qualifications

Elimination of any one of these six items will either seriously affect an inspection program or negate the program.

The inspector will have several books in his library to stay up with the electrical **STANDARDS**.

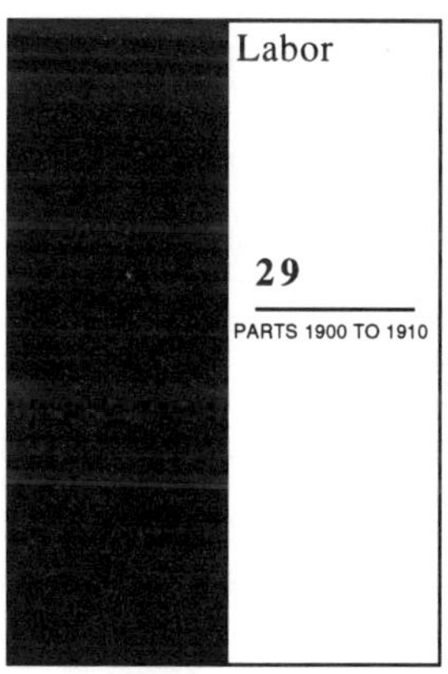

The OSHA law requiring certified electrical equipment in work places is hardly enforced. The flood of electrical equipment; both domestic and foreign, which has never been certified by an approved testing laboratory has placed an unusual burden on the electrical inspector.

The nation's most authoritative and comprehensive document on safety to life in buildings, the code addresses life safety from fire in both new and existing structures. The code is developed by the NFPA Committee on Safety to Life.

Key to determining does the "electric supply station" have an electricity generation or a delivery function (as opposed to solely a utilization wiring function); and is the facility under the exclusive control of, and accessible only to, qualified persons? If the answer is "yes", the **NESC** applies.

The National Electrical Safety Code covers the utility company wiring.

CODES and STANDARDS

Electrical, building, plumbing, mechanical, life safety, safety codes, fire codes, etc. NEMA, UL, ANSI, OSHA, etc. The electrical inspector has a lot of references to read, understand and enforce. The inspector must have a good communication with all the building trades.

From this condensed version of an electrical inspector and his work, it should be obvious that very few people responsible for public safety have such broad duties and responsibilities.

Many electrical inspectors teach the Code in evening vocational schools and apprenticeship training facilities in their area.

SPECIFICATIONS

When planning the construction of a building, an architect is generally hired to prepare the working drawings and specifications covering the details required for bidding and construction.

The electrical specifications for a building are the written descriptions of work and duties required of the architect, engineer, or owner. Together with the working drawings, these specifications form the basis of the contract requirements for the construction of the electrical system for the building.

In general, electrical specifications give the grade of materials to be used on the project and the manner in which the electrical system shall be installed.

The electrical inspector must become familiar with the drawings and specifications, as these construction documents and the Code will form the basis upon which the inspection is made.

During the plan review if errors are detected in the drawings or specifications, they should immediately be reported to those in charge.

Contract specifications prepared by architects and engineers generally require inspection or approval.

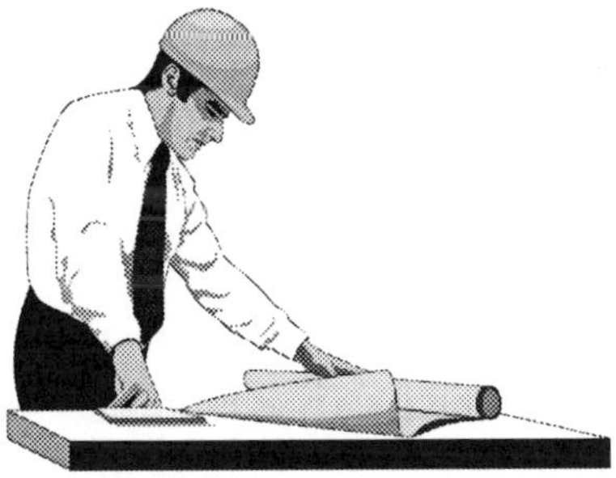

Those inspectors working for architectural and engineering firms must make sure that all work is performed in accordance with the plans and specifications.

Any deviations from plans must be approved.

Responsibilities of the inspector include checking the contractors and electricians working in the jurisdiction for valid licenses, permits, bonding, etc.

The inspector should always be alert for construction work being done in the jurisdiction and provide construction guidelines.

A preliminary conference before construction starts is a good time for the inspector to inform the contractor what the inspector expects in the way of workmanship, performance, drawings, permits, etc. It is also a good time to check for a valid license and the contractor's liability insurance.

A complete review of all drawings and specifications should be as far in advance of the starting of construction as possible. The ability to visualize working drawings comes largely with experience.

It is important to compare the installation with the approved plans. Any changes made from the original approved plans requires approval from the building department.

Once the proper plans and specifications are approved along with an application for the required building permits, the inspection department should promptly review the documents and if everything is okay, issue the permits without delay.

If a new contractor to the area is performing the work, the inspection department should show him courtesy in notifying the contractor of all requirements such as bonding, insurance required, license, local amendments to the Code, etc.

Generally a performance bond is required from the electrical contractor to insure that he will complete the work according to the plans and specifications. Should the contractor default, the company issuing the bond will be responsible for completing the project, usually by hiring another contractor.

The types of insurance that concerns the electrical inspector generally falls into two categories: Liability for injury to other persons or for damage to their propery and loss, damage to, or destruction of the contractor's property or personnel.

Most of the inspector's concern will be with the contractor's liability insurance.

A sample building permit is shown on the next page. The permit is to be posted in plain view during the duration of the construction project.

CITY OF HENRYVILLE BUILDING PERMIT

Building Inspection Bureau --- 871-6635

PERMIT NO's ENGINEERING FLOOR PLAN BUILDING

OWNER ______________________________ DATE ______________

CONTRACTOR ____

ADDRESS ____

This permit conveys no right to occupy any street, alley, or sidewalk, or part thereof, either temporarily or permanently except that specifically provided for in the Building Code. Street or alley grades may be obtained from the City Engineer's office. Depth and location of public sewers may be obtained from the City's Engineer's office. The issuance of this permit does not release the applicant from the conditions of any appliable subdivision restrictions.

Approved plans MUST be retained on job and this card KEPT POSTED until final inspection has been made. The building SHALL NOT BE OCCUPIED until the Certificate of Occupancy has been issued.

SEPARATE PERMITS ARE REQUIRED FOR PLUMBING, GAS, MECHANICAL, SIGNS, ELECTRICAL, SWIMMING POOLS AND FENCE INSTALLATIONS.

Building
FOOTING SLAB/FLOOR LINTEL/COLS INSULATION
FRAMING MISC FINAL

Roof Inspection
DECKING DRY-IN MISC FINAL

Swimming pool
STEEL/GND FINAL

Electrical
TEMPORARY UNDERGROUND ROUGH-IN MISC FINAL

Mechanical
ROUGH-IN MISC FINAL

Plumbing
ROUGH-IN MISC WATER TEST INTERMEDIATE FINAL

Gas
UNDERGROUND ROUGH-IN FINAL

ZONING _____ FINAL ENGINEERING _____ FINAL FIRE _____ FINAL

"FAILURE TO COMPLY WITH THE CONTRACTORS' LIEN LAW CAN RESULT IN THE PROPERTY OWNER PAYING TWICE FOR BUILDING IMPROVEMENTS"

UNLAWFUL TO REMOVE OR DEFACE THIS CARD UNTIL CONSTRUCTION IS COMPLETE

This book was NOT written as a step by step guide of how to make a total electrical inspection, but rather as a guide to make certain the electrical inspector is not overlooking the **CRITICAL AREAS** that have started fires and caused accidents.

As an inspector you have served your time as an electrician and you are aware of the Code rules of installing wiring. There is no substitute for experience. The clamping of conduits and boxes, the stapling of romex, the number of conductors permitted in a box or conduit, a total of 360° of bends in a run, etc.

How many fires were caused because a conduit run had more than 360° of bends? How many people have been injured in a residence because the outlets were 12' 1" on center rather than the required 12'? How many people have been electrocuted because the conduit was not strapped within 3' of the box or the conduit not strapped every 10'?

We can all read our Code book and agree they all are violations. The point this workbook is making; let's also inspect the building for the hazards that are causing damage to the building and injuries to people, so if we have our day in court we can state we have carefully examined the **HAZARDS** within the building.

I know I have found in my travels, inspectors that are overloaded and need to cut corners in order to complete the required inspections for that day. But there are certain items within the building that you can not cut corners on as a competent electrical inspector. That's what this workbook is all about.

The inspector generally works for the building department which is public tax money. It must be remembered that the public, in the end, pays whatever extra cost is caused by requiring safer and better construction, and the public may rightly require a good degree of safety in the construction.

On new construction the electrical inspection is generally in three steps: (1) Temporary power (2) Rough-in (3) Final.

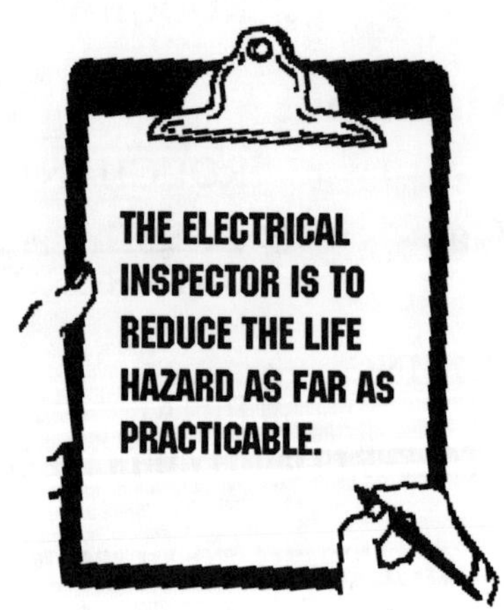

REQUIRED INSPECTIONS

ELECTRICAL

UNDERGROUND INSPECTION:	To be made after trenches or ditches are excavated, conduit or cable installed, and before any backfill is put in place. Code Table 300-5 minimum burial depths will be referenced during this inspection.
ROUGH-IN INSPECTION:	To be made after the roof, framing, fireblocking and bracing is in place and prior to the installation of wall or ceiling membranes. The rough-in inspection must be inspected very carefully as the wiring system within the walls will later be concealed.
FINAL INSPECTION:	To be made after the building is complete, all required electrical fixtures are in place and properly connected or protected, and the structure is ready for occupancy.

THE INSPECTOR

TOOLS

An inspector can't have too many electrical books. I have over 300 electrical books in my library.

- A copy of local code and NEC
- A copy of UL green book
- Flashlight
- Screwdrivers
- Pliers
- Polarity tester
- Tape measure
- Wire gauge
- Torque tools
- Pencil, pen, and paper
- Hard hat
- Continuity Buzzer
- Stop work tags

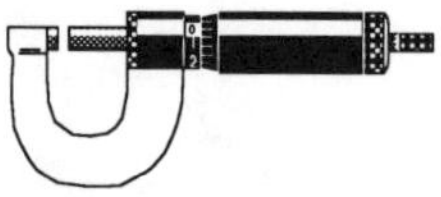

STOP

WORK ON THIS JOB

This building has been inspected and the following correction is required before any further work is done.

Location ____________

Date ____________

DEPARTMENT OF BUILDING AND SAFETY

871-6635

DO NOT REMOVE THIS TAG

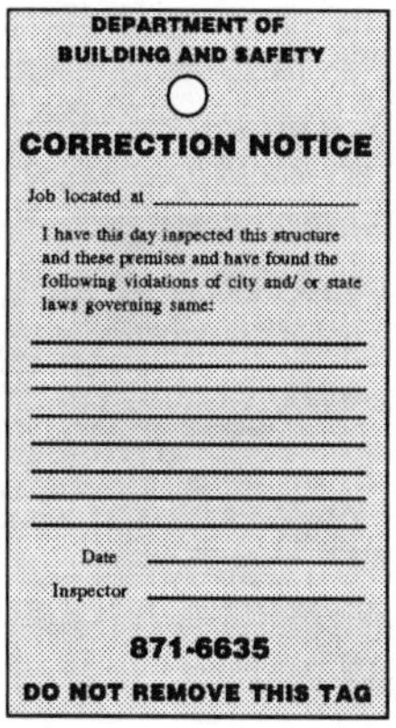

DEPARTMENT OF BUILDING AND SAFETY

CORRECTION NOTICE

Job located at ____________

I have this day inspected this structure and these premises and have found the following violations of city and/ or state laws governing same:

Date ____________

Inspector ____________

871-6635

DO NOT REMOVE THIS TAG

Effective electrical inspections will help to eliminate conditions that lead to electrical shock and fires. The electrical inspector must make sure the electrical installation at least meets the minimum requirements of the National Electrical Code before granting any approval.

A competent and knowledgeable inspector will work with the electrician and explain the intent of the ruling to create a harmonious working environment.

The definition of **AUTHORITY** is an individual cited or appealed to as an **EXPERT**.

If a Code violation should occur, a competent inspector will explain the Code section and the intent of the rule to the electrician. It's called being educative.

In reality we have, in some cases, a **PERSON** filling the slot of electrical inspector which actually downgrades the entire electrical profession. The **PERSON** is the one that states "I'm 90-4, I don't have to quote any Code section, or explain to anyone why it's a violation. It's a violation because I said so, and I'm 90-4 the **AUTHORITY**."

I mentioned previously that **DEFINITIONS** are very important to the inspector. Let's look at these very interesting and fitting definitions.

Remember the definition of **AUTHORITY** is an individual cited or appealed to as an **EXPERT**.

EXPERT. Displaying special skill or knowledge derived from training or experience representing mastery of a particular subject.

EXPERTISE. Expert opinion or commentary.

EXPERTIZE. To give a professional opinion after careful study. To examine and give expert judgment on.

EDUCATIVE. Instructive.

TEACH. To accustom to some action or attitude. To guide the studies of. To impart the knowledge. To instruct by precept, example, or experience. To provide instruction.

If you are the competent inspector then you are the respected one who takes the time to give your expert opinion and to impart your knowledge and provide instruction to help the electrical industry become more educated in the rules of the Code.

Only a person would hide behind 90-4, not an electrical inspector.

SUMMARY

Section 90-4 is the section that refers to the enforcement of the Code rules. The authority having jurisdiction for the enforcement of the Code will have the responsibility for making interpretations of the rules, for deciding upon the approval of equipment and materials, and for granting the special permission contemplated in a number of rules.

The Code may require new products, constructions, or materials that may not yet be available at the time the Code is adopted. In such an event, the authority having jurisdiction may permit the use of products, constructions, or materials that comply with the most previous edition of thc Code adopted by the jurisdiction.

DEFINITIONS

Inspect:	To view closely in critical appraisal.
Inspection:	The act of inspecting.
Inspector:	A person employed to inspect something.
Authority:	An individual cited or appealed to as an expert.
Jurisdiction:	The power, right, or authority to interpret and apply the law.
Qualified Person:	One familiar with the construction and operation of the equipment and hazards involved.
Special Permission:	The **WRITTEN** consent of the authority having jurisdiction.

The difference between a "shall" requirement and a "should" requirement is, in essence, the difference between "possible" and "practical". A "shall" requirement is expected to be met in all conditions. A "should" requirement recognizes that the requirement may not be practical in all cases.

Sometimes the word "minimum" is for clearance or other requirements, the wording generally used is "not less than" to indicate the basic amount that is required for safety purposes.

CHAPTER 2

ROUGH - IN

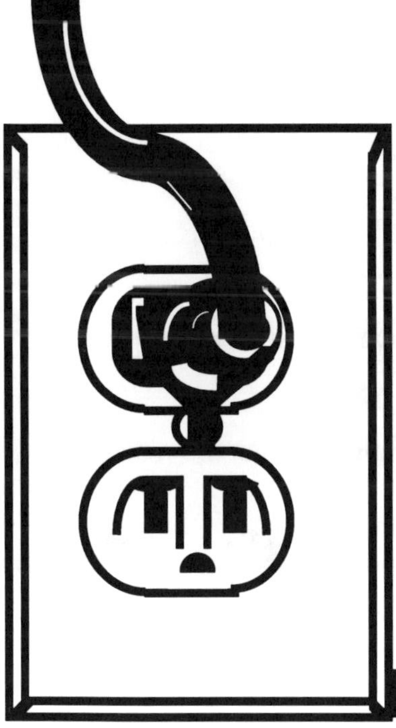

TEMPORARY POWER

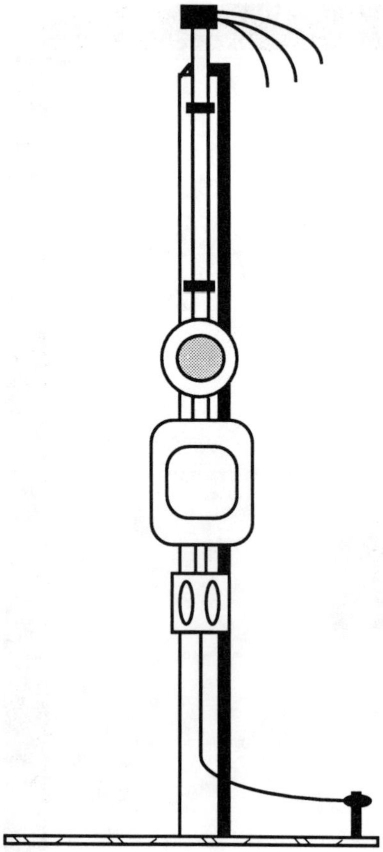

Temporary installations may not encounter the worst weather conditions and for this reason some Code rules are relaxed.

The local utility company generally provides a set of rules pertaining to the installation of a temporary power pole when the electrician requests the meter socket.

The electrical inspector should be familiar with the local utility company's requirements for meter installation.

The National Electrical Code does *not* provide requirements for the sizing or construction of temporary power poles.

Generally the utility company requires at least a 4 x 6 inch size pole as a minimum.

Depending on the location of the pole and the traffic area will decide the minimum clearance from Code section 230-24 (10'-18') to the overhead service drop conductors.

All electrical equipment must be listed and raintight.

Code section 110-26 requires adequate access to the equipment.

All grounding shall conform to Article 250.

Receptacles on construction sites must have ground-fault circuit-interrupter protection per section 305-6.

Receptacles on construction sites shall not be installed on branch circuits that supply temporary lighting.

All lamps for general illumination shall be protected from accidental contact or breakage by a suitable guard.

Flexible cords shall be protected from accidental damage. Where passing through doorways protection is required to avoid damage.

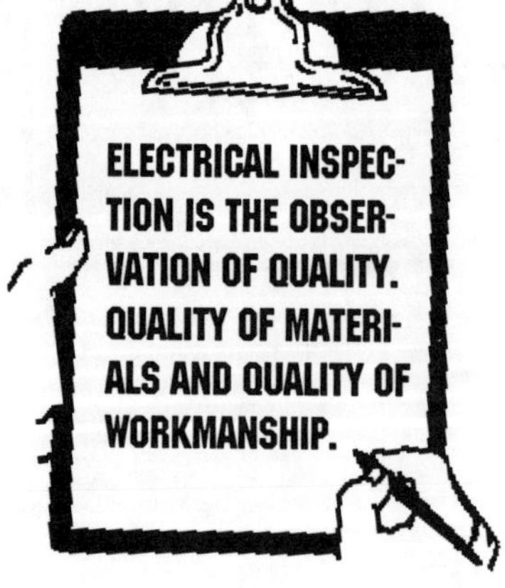

What often happens on construction sites, even though the panelboard on the temporary pole has several branch circuits and receptacles, one cord is run to the building and from this one cord several cords are plugged into a ganged receptacle box. On some jobs *one cord* is running all the drill motors and power saws in the building.

The common use of long cords is frequently the cause of severe burns, eye injuries, and fatalities.

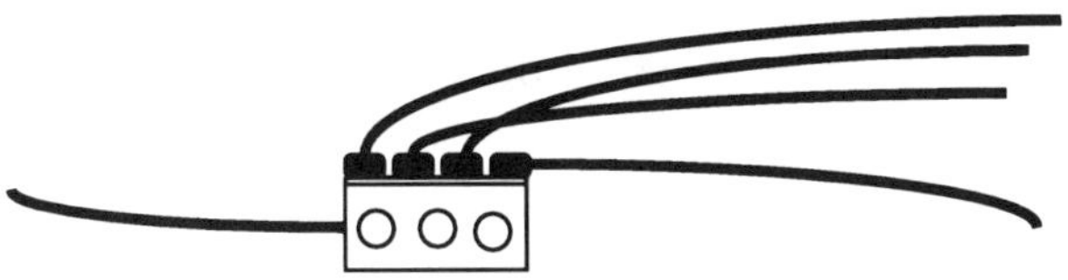

7 1/4" Circular saw

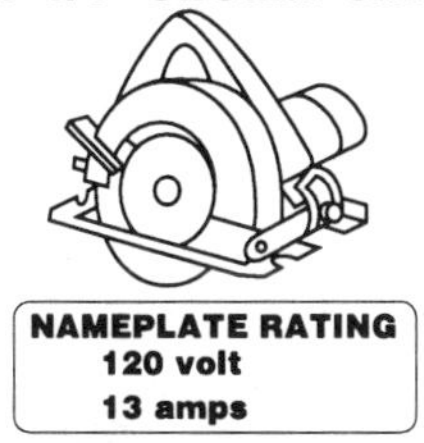

Voltage drop is of extreme safety importance in this situation. Too low of a voltage at the load can cause damage to the equipment and more important, can cause motors to reverse and a saw has been known to reverse backwards over a persons fingers.

#14 AWG ?

#12 AWG ?

#10 AWG ?

Knowing and using the correct size cord for the load is a safety factor that must be addressed on the job.

The electrician as well as the inspector are required to calculate voltage drop in determining the correct cord to be used.

$$\text{Wire Size} = \quad CM = \frac{2 \times 12.9 \times \text{distance} \times \text{load}}{3.6 \text{ volts } (@120\text{v})}$$

Table 8 in the NEC will show the area of circular mils.

The condition of hand power tools and cords should always be observed by the inspector.

Cords are NOT to be installed in raceways unless specifically allowed by the Code.

Cords smaller than #14 are NOT to be spliced.

ROUGH-IN

This chapter is designed to acquaint the electrical inspector with the knowledge of correct inspection techniques for inspecting rough-in wiring methods.

The type of building construction will depend to a great extent on the time the inspection will require. Cable run through studs in a residence, or conduit buried in a concrete pour, must be scheduled more throughly to avoid holding up work on the installation of dry wall or the pouring of concrete. An inspection on a conduit system installed on an exposed wall can be delayed by the inspector as it will not hold up any work on the job.

Generally a residential rough-in inspection of the electrical wiring can be completed in one visit to the site, unless errors are found. Then an additional inspection is required to make certain that the corrections have been made.

On a very large home an inspector can allow a certain portion of the home to be covered before all of the concealed wiring is installed in the other sections of the home.

A residential rough-in electrical inspection would include the following:

- All boxes properly secured to the building structure.
- All concealed wiring should be secured in place.
- If conduit is used in the walls for service or feeder conductors, only the conduit needs inspection.
- Box fill calculations.
- Required receptacle outlets.

Nonmetallic sheathed cable is required by Code to be secured in place at intervals not exceeding 4 1/2' and within 12" of each box or panelboard. Cables that are left not secured in walls are subjected to later damage by the home owner installing a grab or towel bar and screws into the hot conductor of the cable thus energizing the metal bar. This has happened many times to romex cables dangling in the wall. The Code has justifiable reasons for requiring cables to be secured.

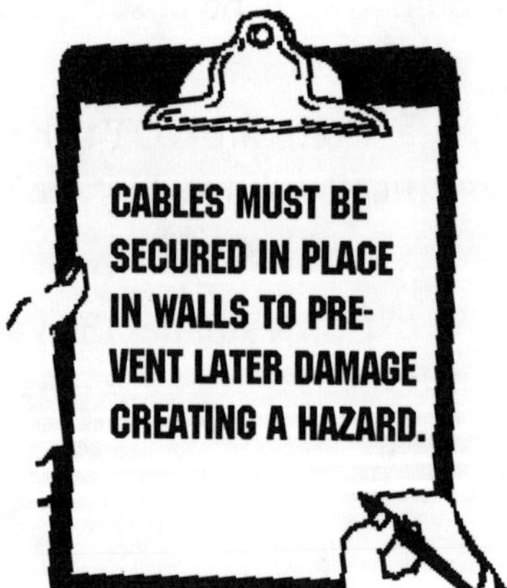

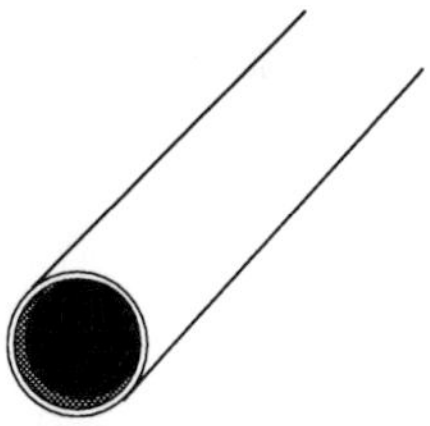

A rough-in inspection on a commercial building follows most of the same guide lines as a residential inspection. The main difference will be the type of wiring method generally will be conduit instead of cable.

The larger size conduits are rarely installed in concrete floor slabs, other than the ground floor slab, because of the possibility of weakening the building structure.

All conduits are to be secured in place, all joints must be cut square, threaded where necessary, reamed smooth, and drawn up tight. All bends in a conduit run must follow the Code requirements for radius, number of bends, etc.

The reason the conduits are to be cut square is in some cases the metal conduit is being used as the equipment grounding conductor to carry the heavy fault currents. If the conduits are not cut square you have less metal contact area with the coupling and heavy arcing can occur at this point.

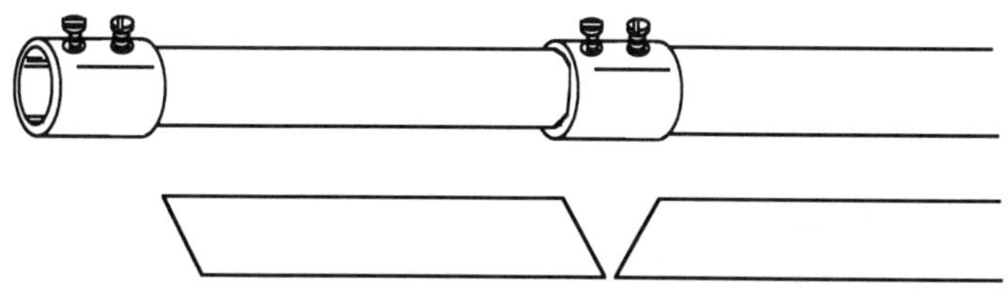

NOT CUT SQUARE

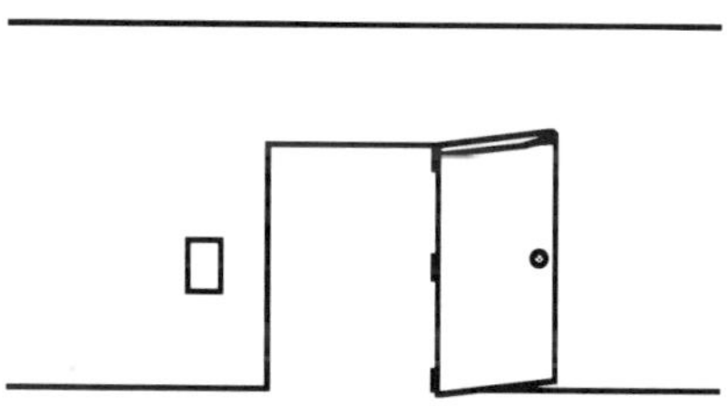

The inspector should always be aware of which way the door is hinged to the swing direction. Boxes for switches are NOT to be installed behind a door. On rough in inspection would be a good time to point out any errors as once the dry wall is installed it becomes more difficult to make the necessary corrections.

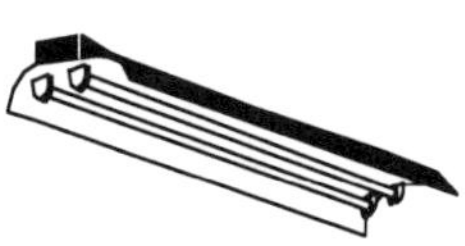

Illumination shall be provided for all working spaces around service equipment, switchboards, panelboards, or motor control centers installed indoors.

The minimum headroom of working spaces around service equipment, switchboards, panelboards, or motor control centers shall be 6 1/2 feet. The exception states this is not required for service equipment or panelboards 200 amps or less in *existing* dwelling units.

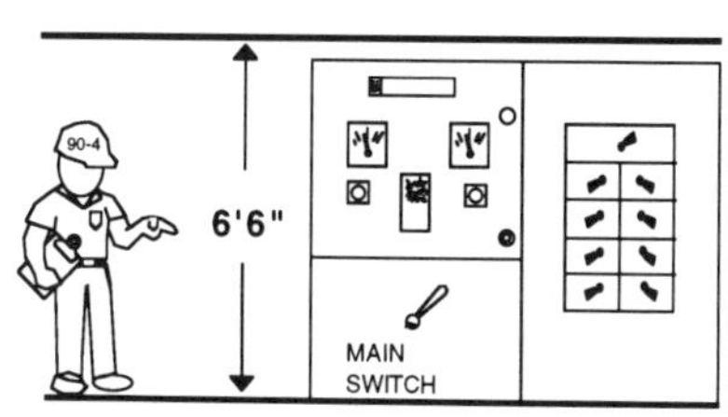

DRILLING AND NOTCHING OF *HORIZONTAL* LUMBER

Drilling holes in the framing lumber weakens the member. But we know holes and notches must be made to route the cable or conduit throughout the building.

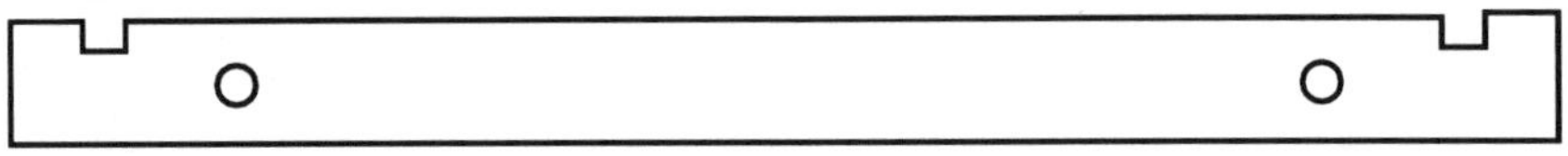

The inspector must always observe the holes being made to assure they conform to the code and they are not causing structual damage to the building.

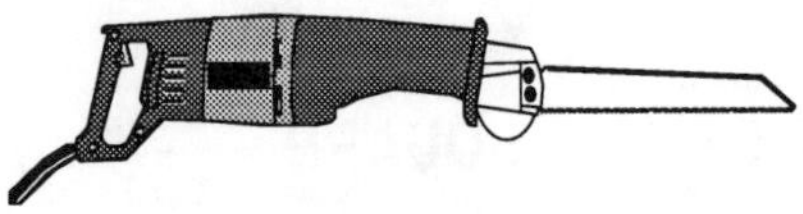

NOTCHING OF LUMBER

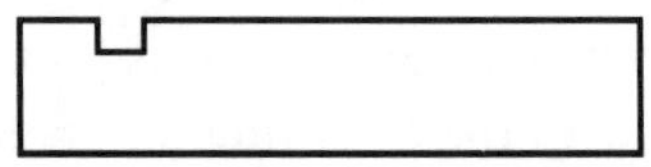

The rules for notching lumber comes from the Building Code. The rules apply to the *repetitive members*. A repetitive framing member such as studs, rafters and joists not over 4" in thickness and spaced not over 24" apart, at least 3 or more in parallel and joined together by a load.

The National Electrical Code in section 300-4a2 permits cable or raceway to be laid in these notches if protected by a steel plate at least 1/16" to protect the cable or raceway from the penetration of nails or screws.

Steel plate

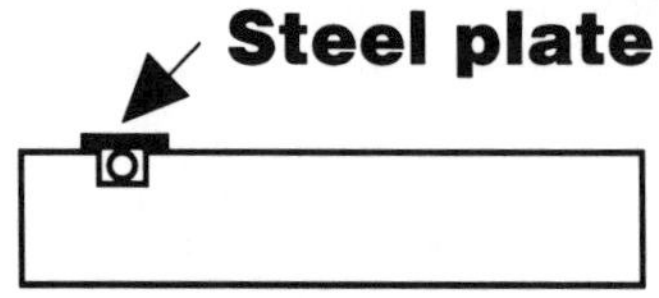

These floor joists are an example of repetitive members as there are at least three, they are 2 x 8 's, 16" on center, joined by a sheet of plywood to distribute the load.

To cut into a single timber or beam it must be detailed and approved on the drawing.

Do not cut into a timber or beam

The depth of the notch is allowed up to 1/6 of the depth of the member. A 2 x 6 has a measured depth of 5 1/4" x 1/6 = approximately 7/8" maximum depth of a notch.

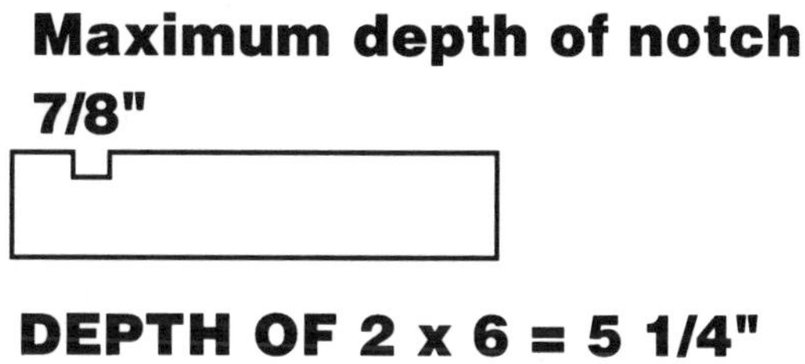

No notches are permitted in the *center 1/3* of the member, only in the outer 1/3.

A good rule to follow is the notch length should not be more than 1/3 the depth of the member. And only one notch at each end (outer 1/3) of the member.

A 2 x 6 with a measured depth of 5 1/4" x 1/3 = 1 3/4" allowed length of notch.

*Bored holes are permitted in the center 1/3 if, NOT within 2" of the top or bottom and NOT to exceed 1/3 the depth of the joist.

A member has a compression side and a tension side. The compression side is the side where the load is on. The tension side is the under side.

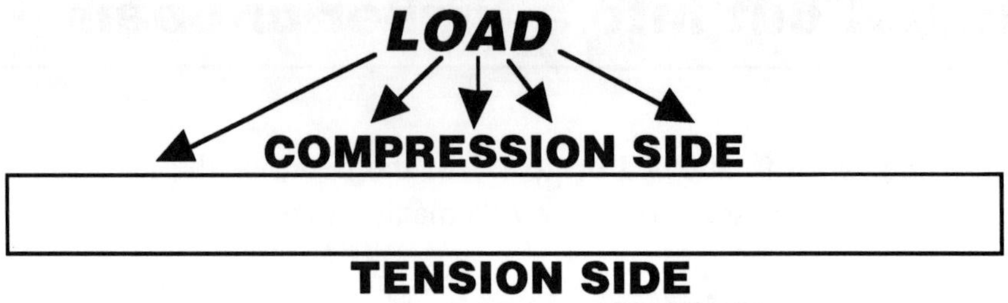

As the load is applied the tension is shown on the bottom side.

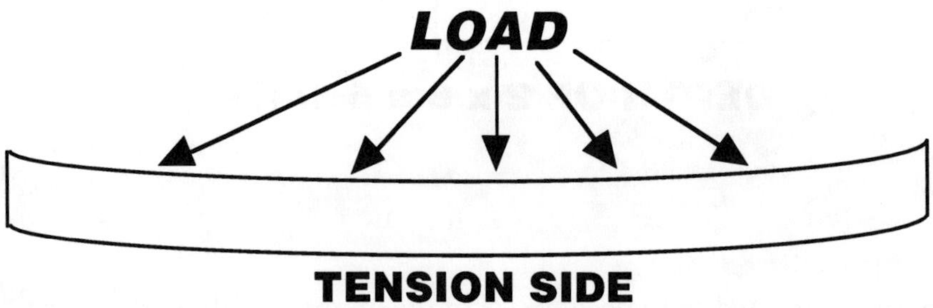

Notching is permitted on both the compression and tension side of 2-by's or 3-by's, but *NOT* 4-by's.

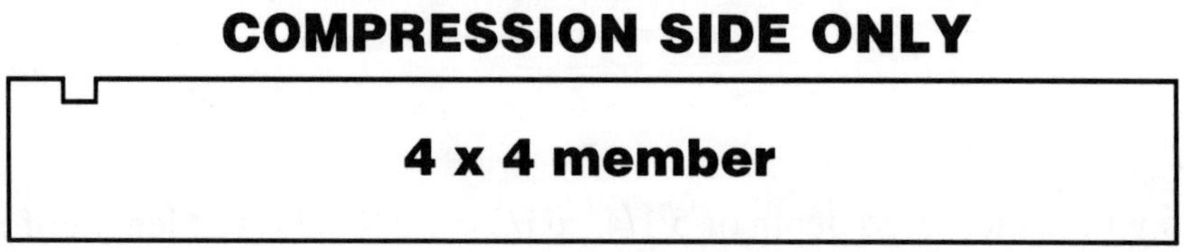

Notches are permitted on each end of a 4-by on the tension side where a support occurs.

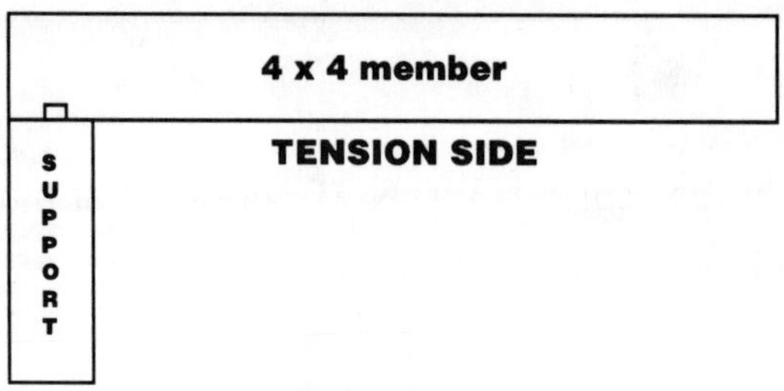

Notches at the end of members are permitted to be 1/4 of the depth, deeper than the 1/6. Example: A 2 x 6 that measures 5 1/4" in depth a 1/4 = a notch depth of 1 1/4".

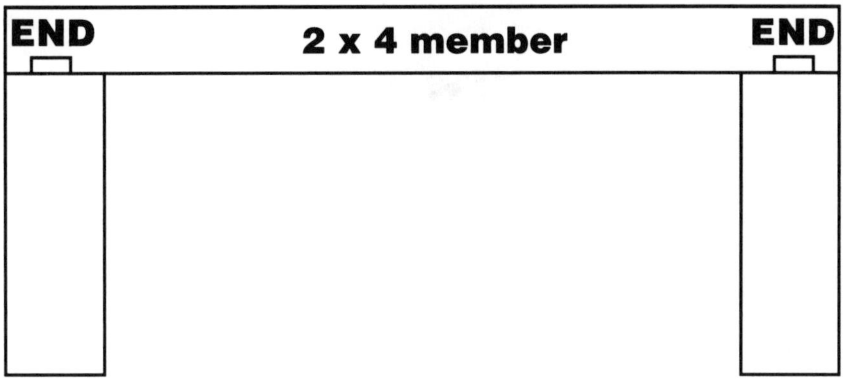

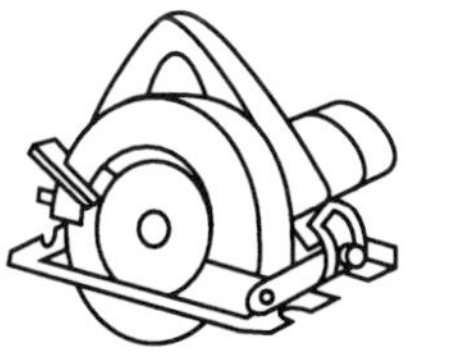

The cutting depth of the saw blade is considered into the maximum depth of the notch. Don't set the saw blade deeper than the maximum depth permitted.

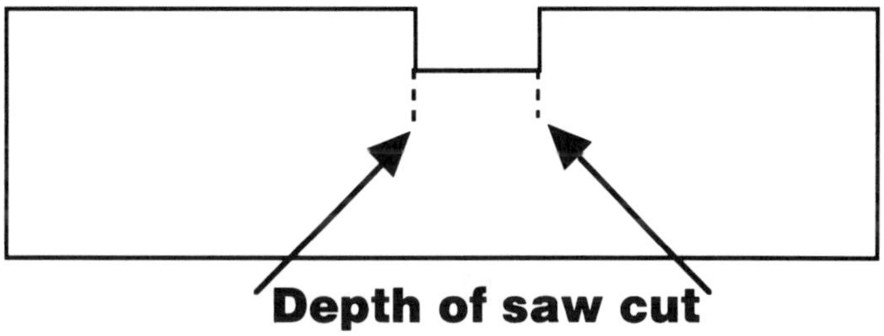

DRILLING OF JOISTS

Holes drilled in joists shall NOT be within 2" of the top or bottom of the joist, and the diameter of the hole shall not exceed 1/3 the depth of the joist.

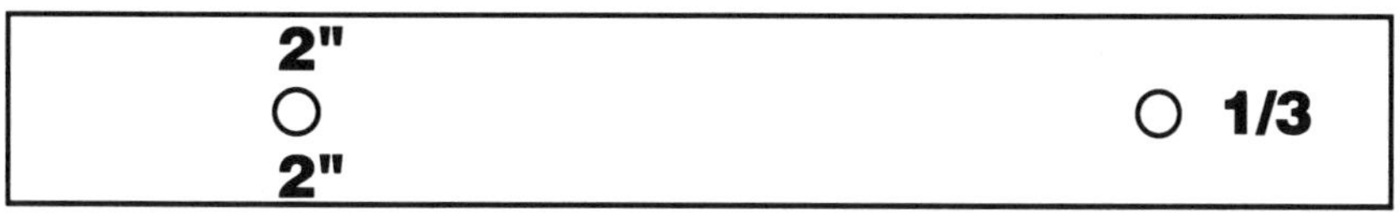

FLOOR JOIST

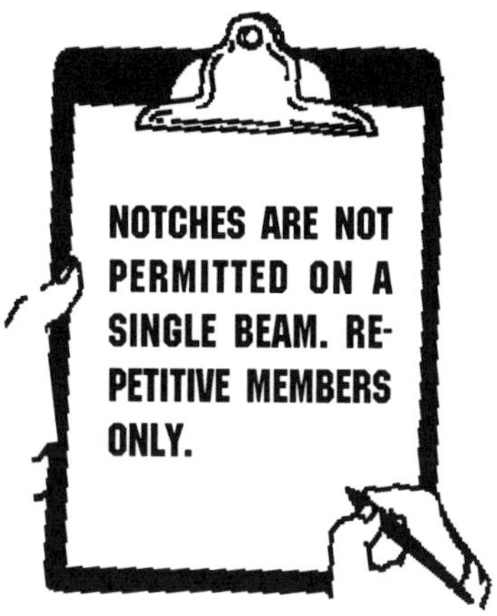

DRILLING & NOTCHING VERTICAL FRAMING MEMBERS (STUDS)

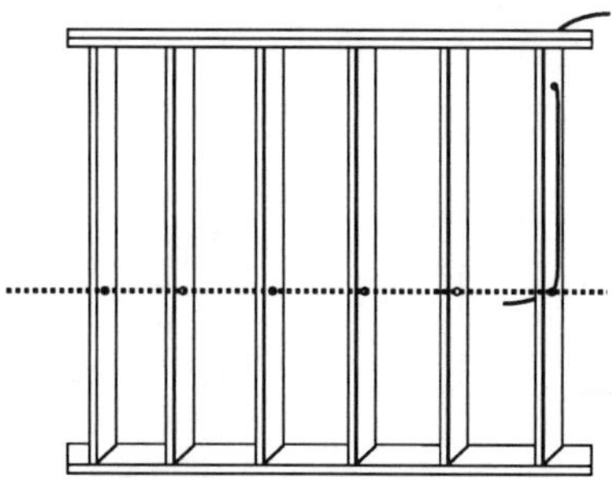

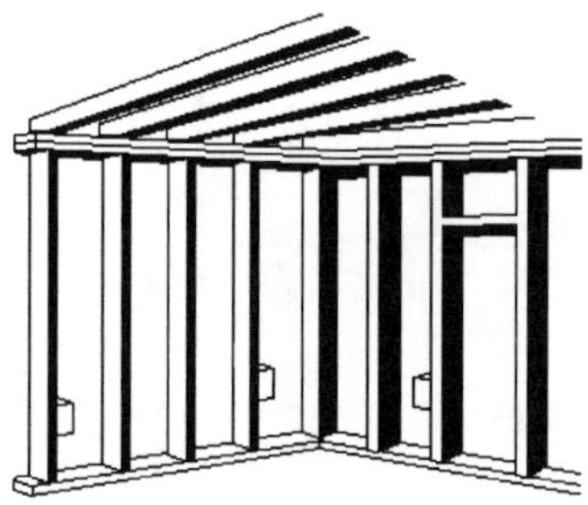

NOTCHING OF STUDS

The Building Code permits the notching of any stud up to 25% of its width. A standard 2x4 stud has a width of 3 1/2" x 25% = 7/8" notch is permitted in any vertical 2 x 4 stud.

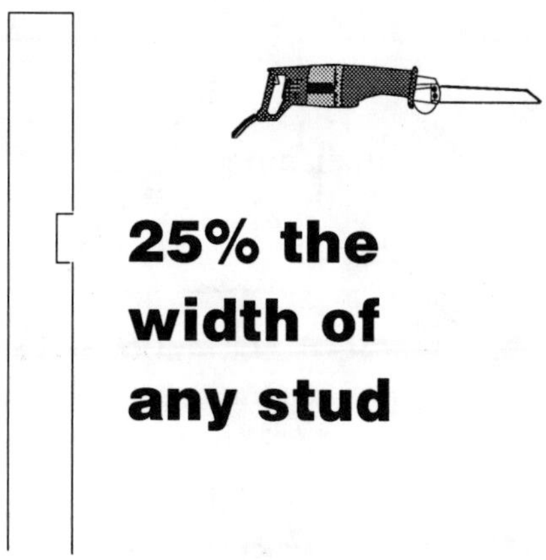

The notch size can be increased to 40% the depth of a width on a *nonload* bearing wall. 3 1/2" x 40% = 1 3/8" notch.

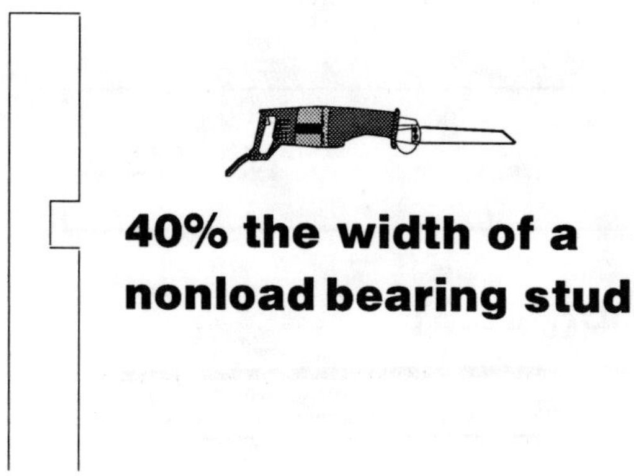

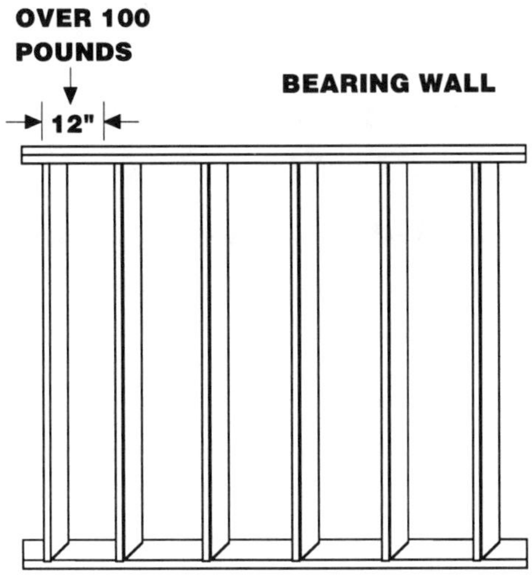

A *load bearing* wood stud wall is a wall which supports more than 100 pounds per lineal foot of superimposed load.

One building code defines a *bearing* wall as a wall supporting any vertical load in addition to its own weight.

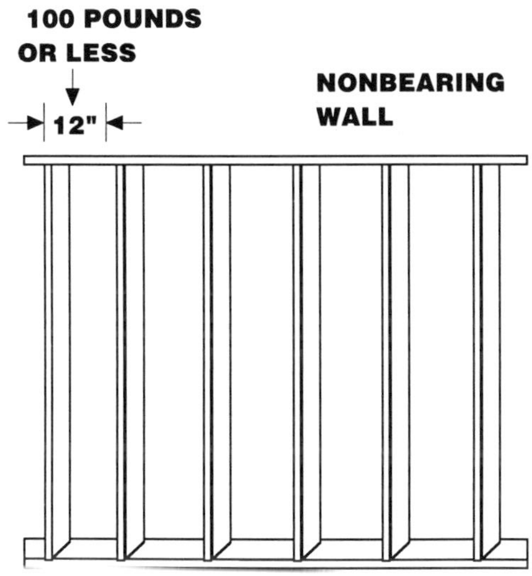

A *nonload bearing* wood stud wall is a wall which supports 100 pounds or less per lineal foot of superimposed load. (It would NOT support any weight of a ceiling).

One building code defines a *nonbearing* wall as a wall which supports no vertical load other than its own weight.

DRILLING OF WOOD STUDS

The Building Code permits a hole not greater than 40% of the stud width in any wood stud (load bearing or nonload bearing). However, the hole is not permitted within 5/8" to the edge of the stud. Holes and notches are NOT permitted at the same location in a stud. A hole closer than 5/8" to the edge of a stud will be considered a notch.

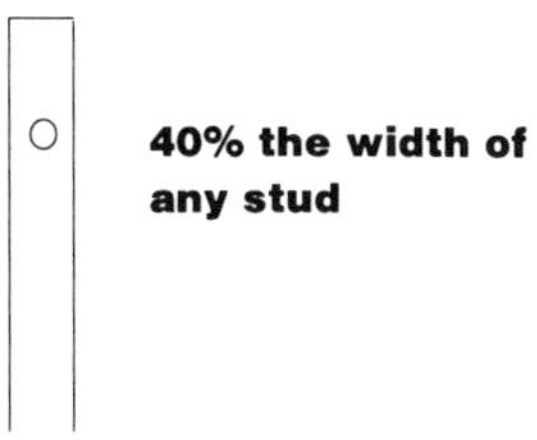

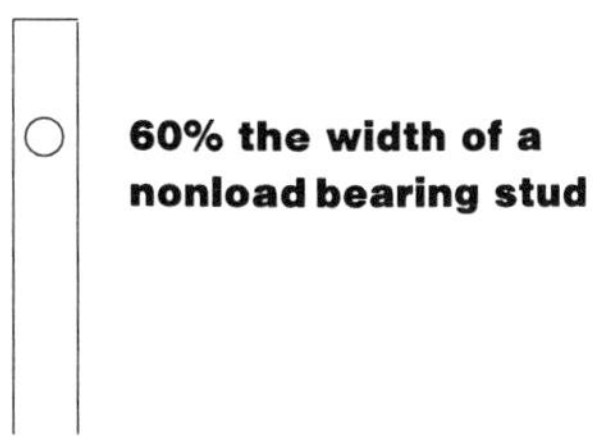

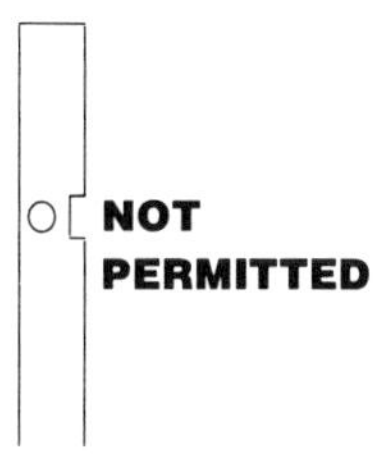

DRILLING AND NOTCHING

A hole of up to 1 3/8" is permitted in any 2x4 stud

A hole of up to 2 1/16" is permitted in a 2x4 stud in a nonload bearing wall

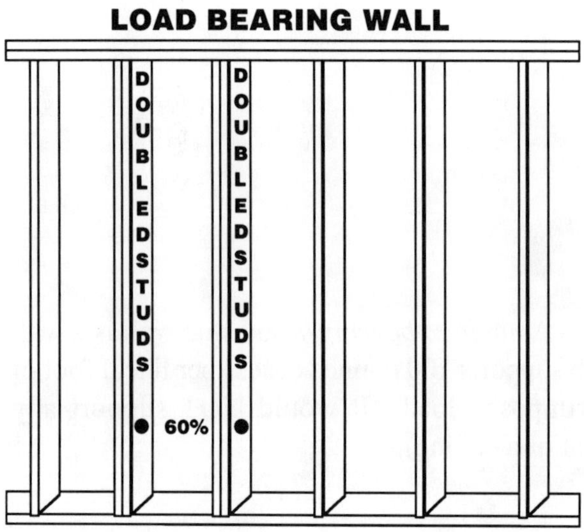

A 60% of the width hole can be drilled in a *load bearing* wall if each drilled stud is *doubled*, provided NOT more than *two* successive doubled studs are drilled.

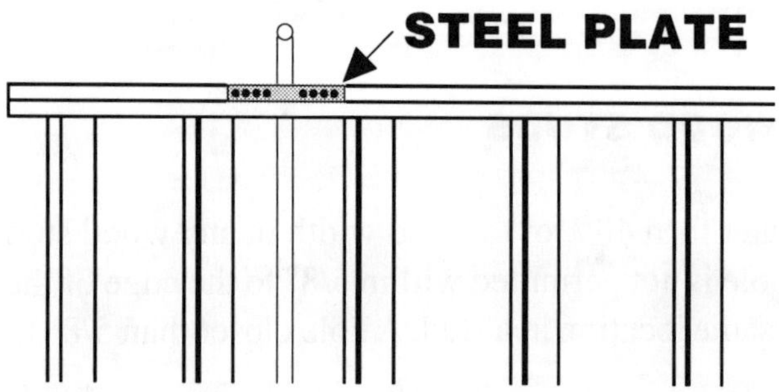

When drilling or cutting into the top plate is required, a steel plate measuring at least 1/8" in thickness and 1 1/2" in width shall be fastened across the opening with not less than four 16 penny nails.

NAIL PLATES

The National Electrical Code in section 300-4a2 permits cable or raceway to be laid in these notches if protected by a steel plate at least 1/16" to protect the cable or raceway from the penetration of nails or screws.

When holes are bored in the framing members, the holes must be kept 1 1/4 inches from the nearest edge. This is to prevent nails or screws from being driven into the cable when the paneling or drywall material is being installed. If this clearance of 1 1/4 inches isn't possible, then metal nail plates must be placed on the outer edge to protect the cable. A standard 2" x 4" stud measures 3 1/2" on the 4" side, if the hole is drilled in the center of the stud, nail plates may not be necessary. Closer to either outer edge will require nail plates.

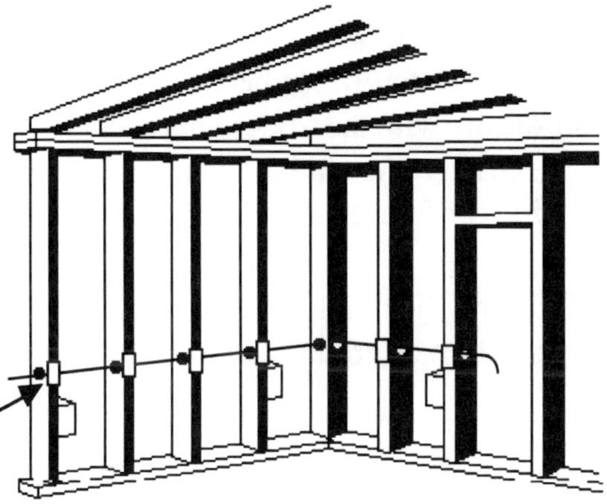

When metal studs are used, bushings or grommets are installed in the holes to protect the cable from the sharp edges.

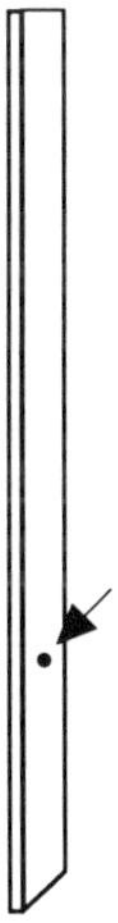

A 2x4 stud is 3 1/2" in width. The code requires a 1 1/4" clearance from the nearest edge. From each edge in 1 1/4" would permit a maximum 1" diameter hole in the center of a 2x4 stud.

A hole larger than 1" diameter would require a metal nail plate for protection.

Truss members and components shall not be cut, notched or drilled, spliced or otherwise altered in any way without written concurrence and approval of the design engineer.

No additional loading of any member, such as HVAC equipment, water heater, shall be permitted without such additional load being incorporated in the engineering design.

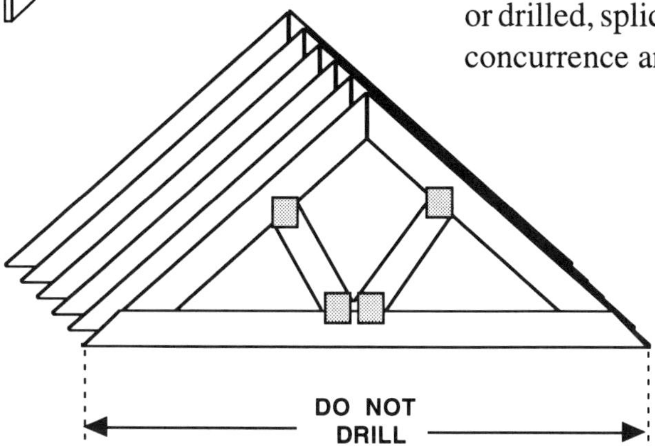

NAIL PLATES

INSPECTION OF BOXES AND RACEWAYS

BOXES

The Code contains provisions considered necessary for safety. Compliance therewith and proper maintenance will result in an installation essentially free from hazard but **NOT** necessarily efficient, convenient, or adequate for good service or **FUTURE EXPANSION** of electrical use.

Hazards often occur because of overloading of wiring systems by methods or usage NOT in conformity with the Code. This occurs because initial wiring did not provide for increases in the use of electricity. An initial adequate installation and reasonable provisions for system changes will provide for future increases in the use of electricity.

Limiting the number of wires and circuits in a single enclosure will minimize the effects from a short-circuit or ground fault in one circuit.

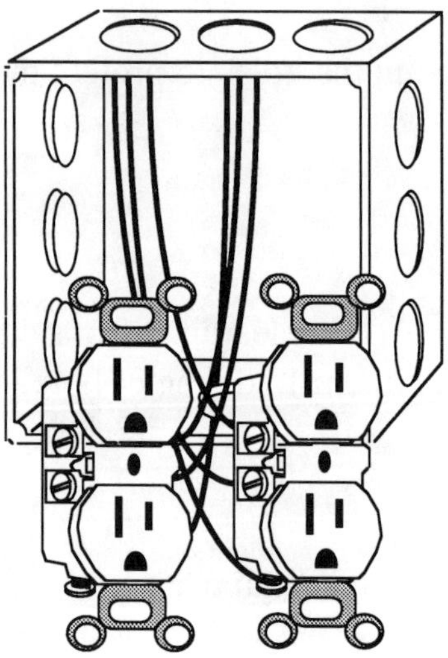

Installing more than the Code maximum permitted conductors in a box has caused accidents and fires. Forcing the receptacles into a crowded box tends to loosen the connections thus resulting in a high resistance fault (arcing heat).

Boxes shall be of sufficient size to provide free space for all enclosed conductors. In no case shall the volume of the box be less than the fill calculation.

Proper calculations in box fill will result in conductors not being jammed into the box causing nicks or damage to the conductor insulation resulting in possible grounds or short circuits.

A very dangerous fault is a **HIGH-RESISTANCE FAULT**(loose connection). The problem with this fault is due to the high resistance the overcurrent device doesn't receive enough current to open the circuit, thus heating continues and then **ARCING** and then the fire starts.

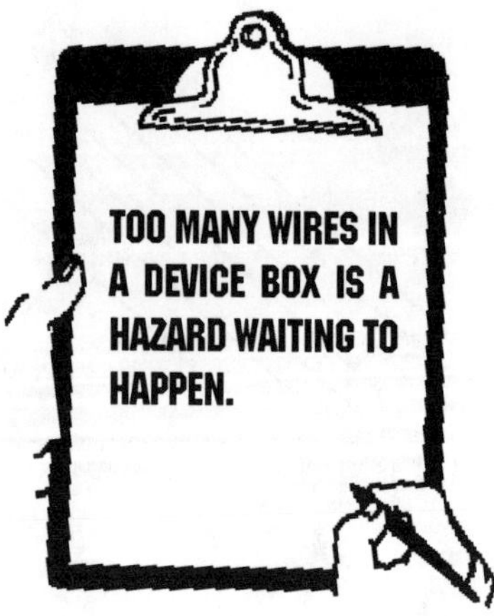

The Code requires all devices and splices to be enclosed in a box for safety. The box also provides a mechanical support for the device. Boxes are made of metallic or nonmetallic material.

There are two kinds of boxes. A device box and a ceiling box. The device box is designed to hold the receptacles and switches. The ceiling box is for installing light fixtures and fans.

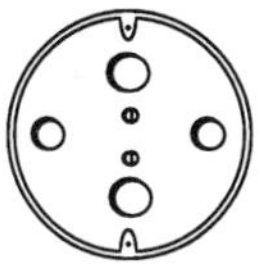

There are three types of ceiling boxes: Octagonal, round, and square. The octagonal is the most used. The round box is sometimes called the *pancake* box.

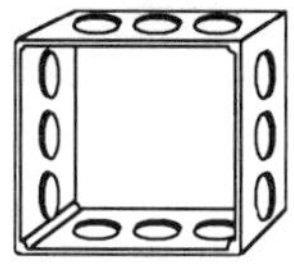

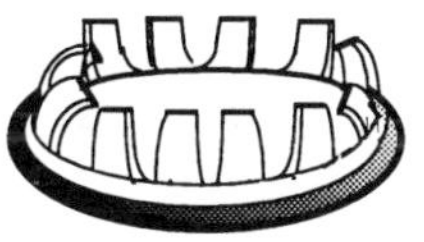

The Code requires that all unused openings in a box be covered. A hole cover is used to plug an unused opening in a box. The hole cover snaps in place and can be removed simply by knocking it back out.

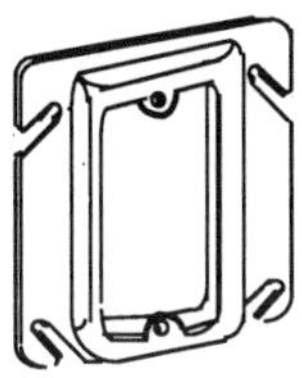

Box extensions often called "*plaster rings*" are added to the box to give it more cubic inch capacity where needed. Basically, a box extension is a box with the back removed. They are available in different depths.

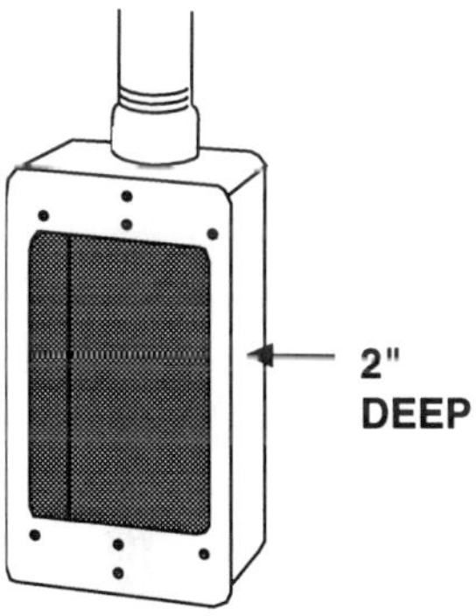

FS BOX

FS and FD boxes provide excellent service in arcas where boxes are subject to *rough usage*. They can serve as pull boxes or used as device boxes.

FS and FD cast device boxes are made of malleable iron, aluminum for use with *threaded* rigid conduit and IMC.

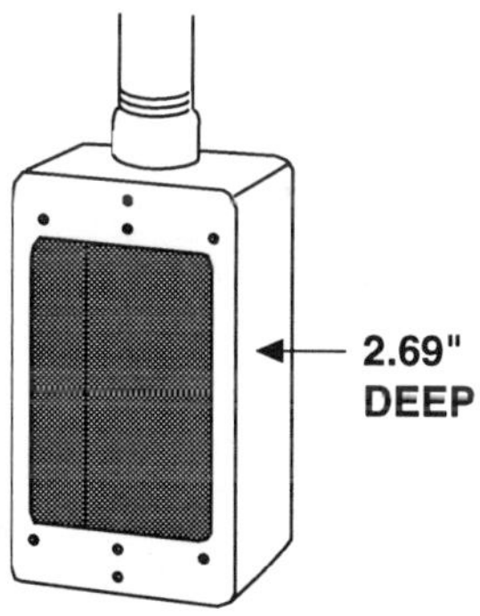

FD BOX

They are corrosion-resistant which makes them ideal for indoor or *outdoor* installations. They are weatherproof, raintight, and dust-tight when used with cast gasketed covers. They will accommodate the standard flush wiring devices.

Boxes are to be installed so the wiring is accessible. There is an exception if a **LISTED** box is used outdoors it can be covered by gravel, light aggregate, or light soil if the location of the box is **IDENTIFIED** and can easily be dug up.

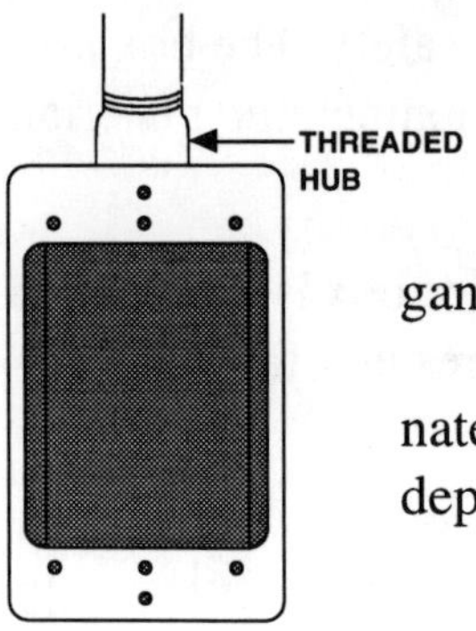

FS and FD boxes are available in single, two and three gang, and tandem.

The "F" is designated for *fitting* box. The "S" is designated for *shallow* in depth. The "D" is designated for *deep* in depth.

The boxes are either nailed or screwed to the vertical studs of the wall. There is no Code rule on the mounting height of the box from the floor, except the receptacle cannot be over 5' 6" above the floor and still count as a required receptacle in the spacing requirements.

Check the approved plans as they may indicate height of boxes to meet handicap requirements.

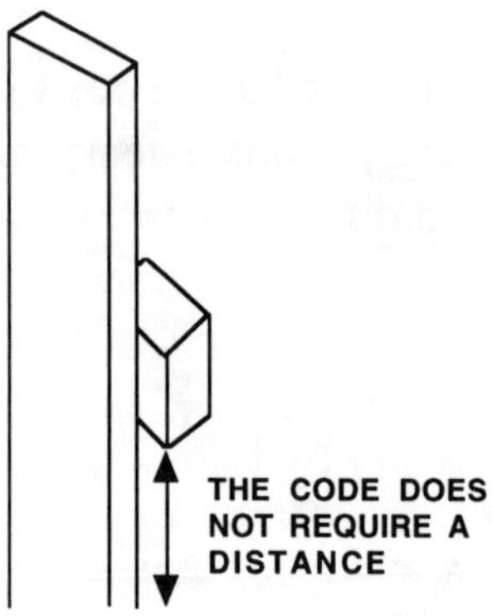

Generally, receptacle boxes are mounted 12 - 16" above the floor.

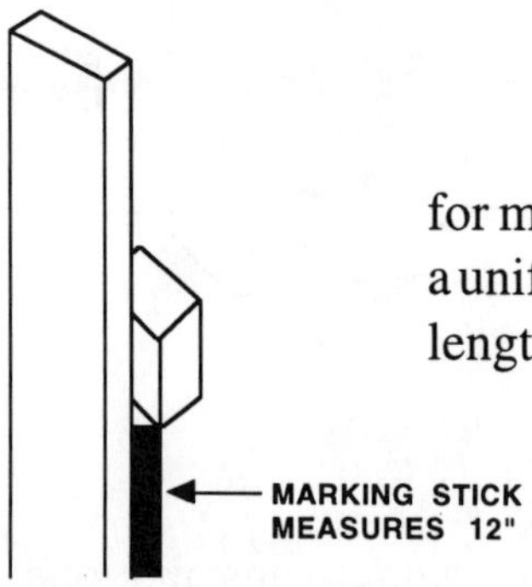

A marking stick cut from a scrap piece of lumber is handy for marking each receptacle box location as it will keep them all in a uniformed height from the floor. Often the electrician will use the length of a hammer to mark the height of each receptacle box.

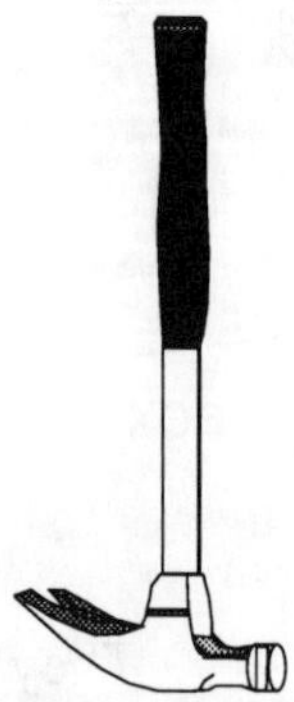

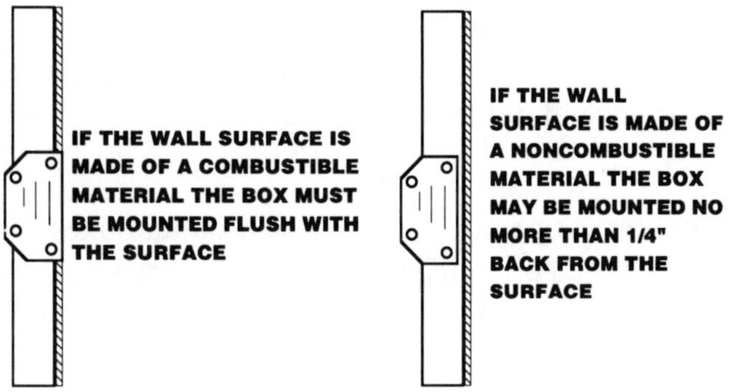

If the surface of the wall is to be of a combustible material such as wood, the boxes for outlets must be mounted flush with the finished surface. If the surface of the wall is to be of a noncombustible material such as dry wall, the boxes for outlets may be recessed in the wall not more than 1/4".

Some drawings show switch box mounting heights at 48" from the floor. Being an electrician over the years has caused me to disagree with this 48" measurement in some cases. When drywall is installed at 48" in height and 12 feet horizontally, the box will be located in the taped joint at 48". Same condition exists with the wall that has 4' x 8' paneling laid horizontally as a wainscot for protection of the lower half of the wall, the top trim molding will cover the box at 48". The bottom of the box should be at least 52" from the floor to stay clear of the 48" of the material being installed, otherwise the joint will fall at the box location. My experience has been when the joint compound is removed from the boxes the wire insulation gets damaged in the process.

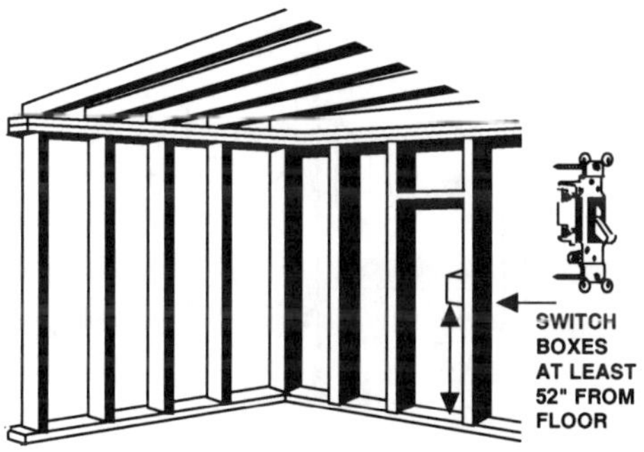

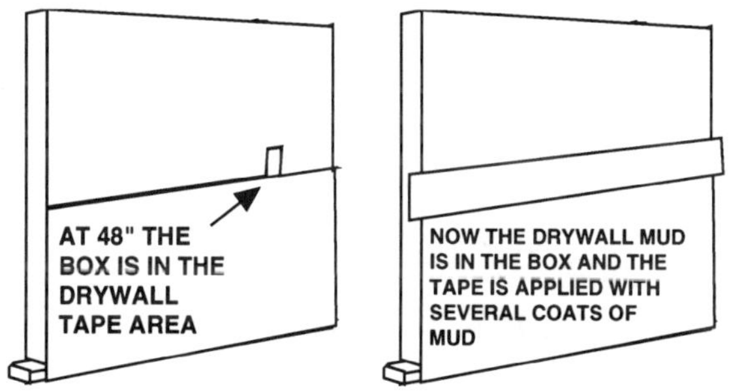

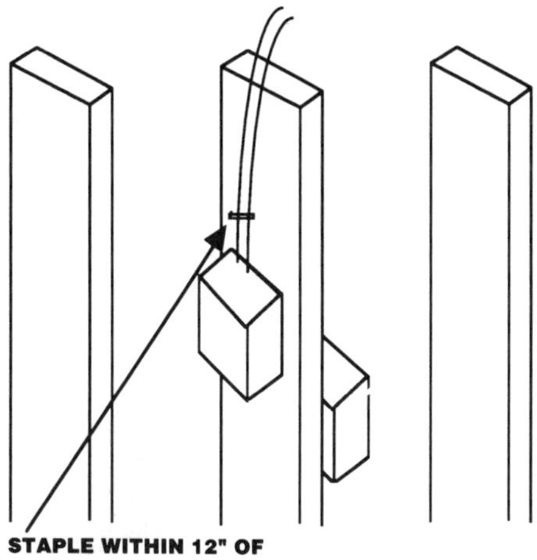

The cable must be stapled within 8" of the box if it is not clamped to the box or within 12" if it is clamped to the box, and must be stapled at no more than 4 1/2 feet of cable run. To avoid damage to the cable, the staples should not be driven too tightly against the cable. If the cable is ran through holes such as through the studs from box to box, it is considered supported and stapling is not necessary.

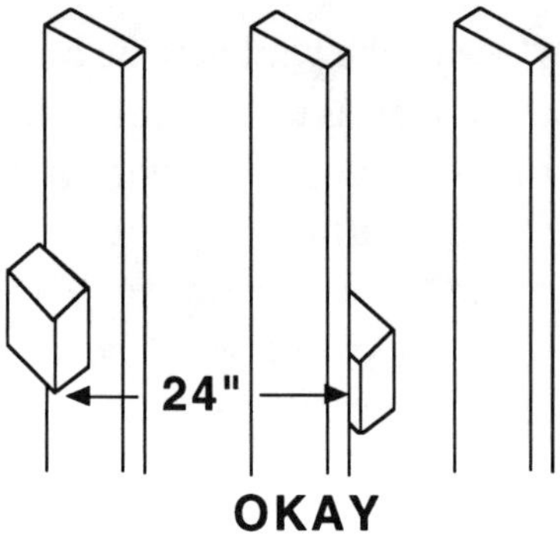

The Building Code requires outlet boxes on opposite sides of the fire resistant wall or the shaft enclosure shall be separated by a horizontal distance of not less than 24".

The exception is openings for electrical outlet boxes of any material are permitted provided such boxes are listed for use in fire resistant assemblies and are installed in accordance with their listings.

Back to back boxes or boxes in the same stud chase to opposite rooms are not permitted in fire rated walls. A good example would be a hotel room where a fire in one room could spread to an adjoining room through the boxes even though the wall is fire rated. Some codes permit boxes if in a different stud chase.

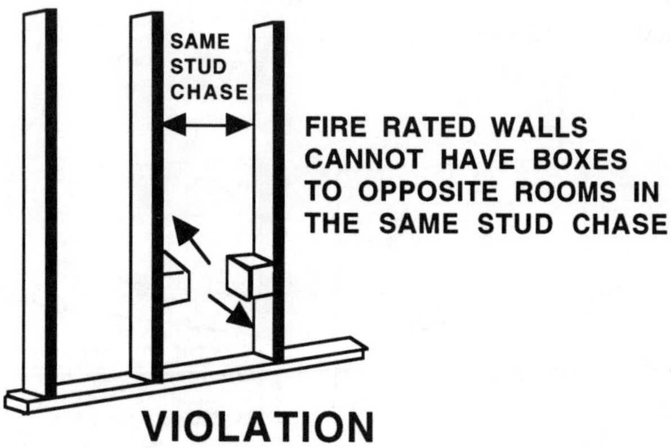

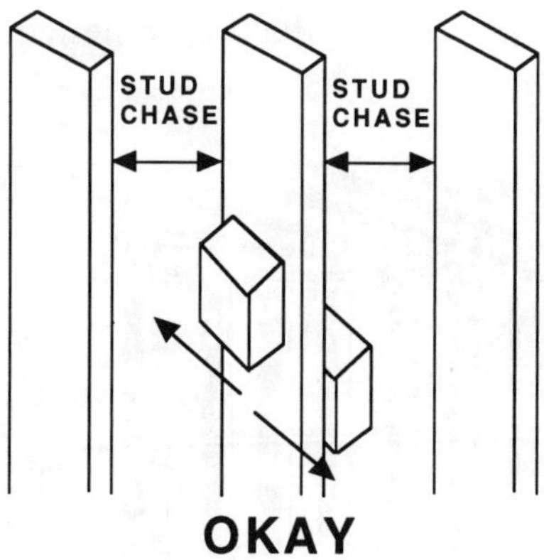

BOX SUPPORTS

Code section 370-23 requires a box to be rigidly and securely fastened in place.

The violation I see everywhere in my travels is the violation of section 370-23e. Boxes that contain **DEVICES** and are not over 100 cubic inches in size and have threaded entries or hubs shall be considered adequately supported where two or more **CONDUITS ARE THREADED** wrenchtight into the box hubs and each conduit is supported within 18" of the box.

This section says "where the **CONDUITS** are threaded wrenchtight". It doesn't say where the *fittings* are threaded wrenchtight. Rigid metal or intermediate metal conduit is threaded, not PVC.

As I walk through city parks and other outdoor areas that have receptacles it's very common to see the box supported by two PVC conduits from below grade. In some cases I've even seen a single PVC conduit supporting the box.

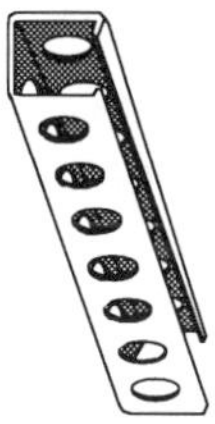

Below grade PVC is the best choice to use for soil conditions. But the box is NOT considered supported by this type of raceway. To meet the Code the *box* would have to be supported by another means such as bolting it to a piece of kindorf channel.

Code section 370-27c states that outlet boxes shall NOT be used as the sole support for ceiling (paddle) fans unless the box is listed for the application as the sole means of support.

I have been asked where in the electrical Code does it give the minimum height for a paddle fan?

This is not found in the electrical Code. The **MECHANICAL CODE** states: Rotary fans without fan blade protection shall provide for not less than 8' of clearance from the finished floor level to the bottom side of the unprotected fan blades.

The exception is for fan blades of low speed residential type ceiling fans installed *within dwelling units* shall be located at least 6' 8" from the finished floor.

DWELLING UNITS
6' 8"

COMMERCIAL
8'

RACEWAYS

Conduit provides mechanical protection to protect conductors, permits easy wiring modifications.

If a metal raceway is being used as the equipment grounding conductor, bending the conduit is better than using several condulet fittings. Each joint is a possible loose connection in the required effective grounding path to carry the heavy fault currents.

Many electricians assume that the conduit is used essentially for mechanical protection and do not realize the likelihood of *sparking* when the conduit is called upon to carry fault current.

Where metal raceway is used as the equipment grounding path, couplings, bushings, setscrews, and locknuts should be checked to see that they are tight and properly seated. Any metal raceway used as the equipment grounding path should be examined carefully for rigid mounting and secure joints. Metal conduit is part of the electrical circuit, therefore the joints must be tight and clean to carry the large current under accidental fault conditions.

The Code places no restriction on the *size* or *length* of metal conduit. It must have continuity as well as the required conductivity to pass enough current to open the breaker or blow the fuse under a fault condition.

Joints in conduits and raceways require *special* consideration to avoid a shower of sparks and accompanies a fire hazard during a fault condition.

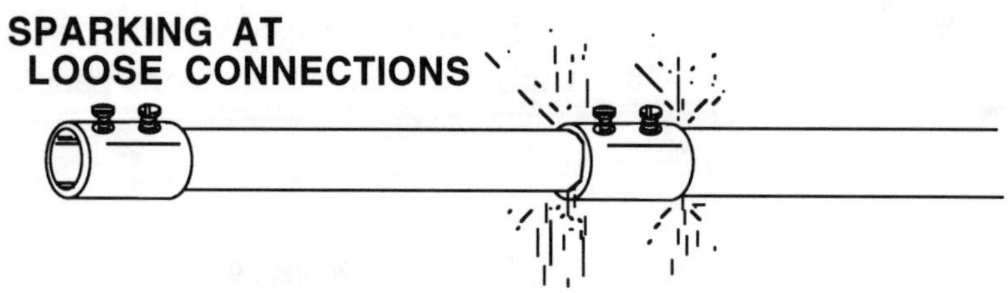

Some electricians know better but are careless in their work and this results in loosely assembled conduit systems which will definitely throw sparks if they are called upon to carry fault currents.

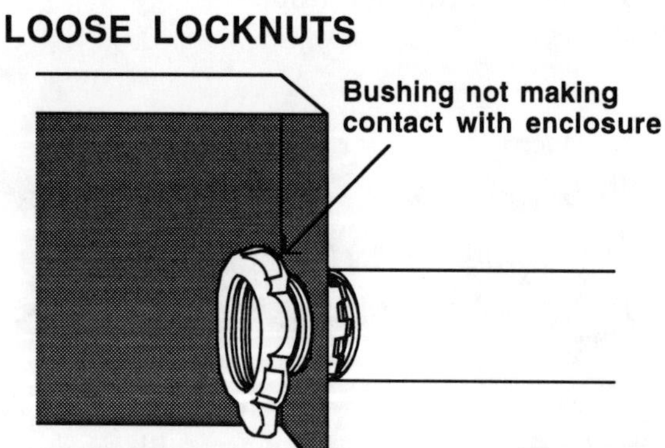

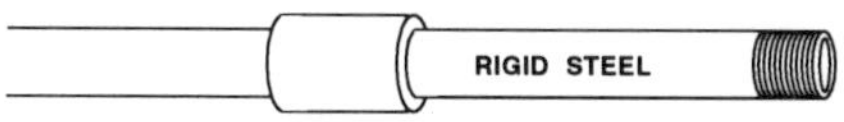

Rigid steel conduit must have the threads securely tightened.

A high-current test on rigid steel conduit produced a shower of sparks from about half the couplings in the conduit run. From one came a blowtorch stream of sparks which burned out many of the threads. Several small fires set in nearby combustible material would have been serious if not properly extinguished. The conduit had been installed by a crew regularly engaged in such work and they gave assurance that the joints had been pulled up to normal tightness and perhaps even a little more.

The importance of proper equipment grounding becomes greater with the larger size feeder circuits and the availability of higher short-circuit currents.

There is no practical substitute for a well designed and well assembled conduit system.

The Code says an equipment grounding conductor can be insulated or bare. If a *bare* conductor is used in a *metal* conduit there can be arcing damage during a ground fault condition. The arcing could cause damage to the other insulated conductors in the conduit. Using an insulated conductor in a metal conduit is the best choice, even if not required. *This would be a good rule to put in your local code.*

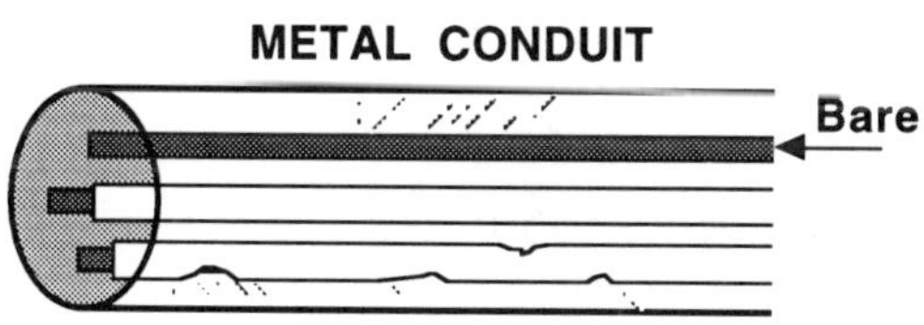

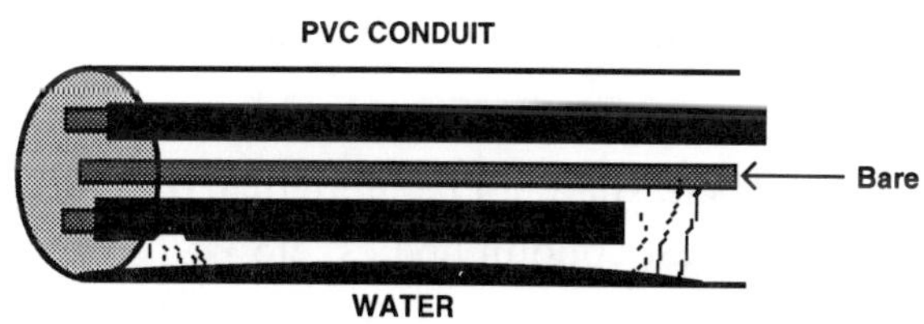

In some cases a designer will require a *bare* equipment grounding conductor to be installed in PVC conduit when the conduit is installed below grade.

Almost all underground PVC conduits will have water seep into them over a period of time. Because of this, the conductors will have a direct connection to ground. This contact to ground is not a problem in itself, but the problem is this: If the insulation in one of the ungrounded (hot) conductors breaks down, cracks or gets nicked during installation, there will be no return path to the panel in which the circuit originates if the grounding wire is *insulated.* This means the leakage current from the hot wire to the water will run through the ground rather than returning through the grounding conductor to open the breaker. The *bare* equipment grounding conductor will make contact with the water in the conduit and carry the fault current to trip the breaker.

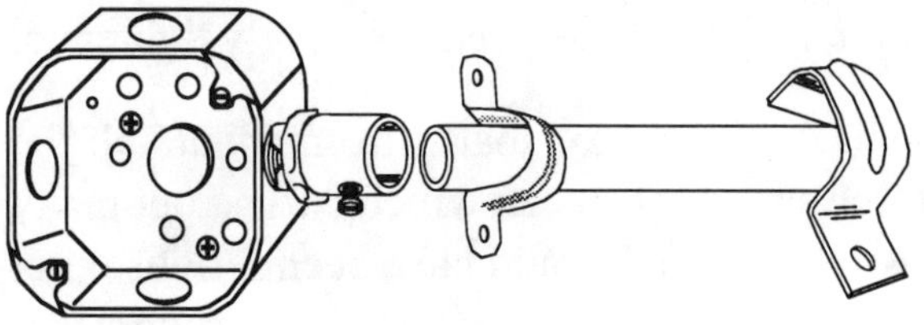

Conduit is to be clamped within 3 feet of a box and every 10 feet in length.

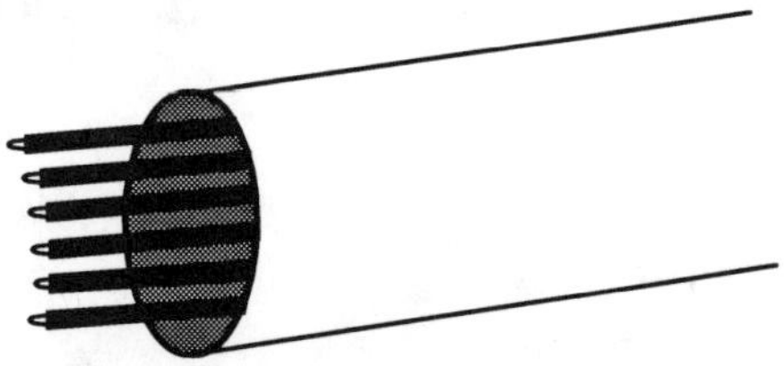

The number and size of conductors in any raceway shall not be more than will permit dissipation of the heat and ready installation and withdrawal of the conductors without damage to the conductors or to their insulation.

Raceways shall be installed complete between outlet, junction, or splicing points **prior** to the installation of conductors per Code section 300-18.

Raceways must be installed so as to substantially maintain the *integrity* of the fire rated construction of the building. All rules relating to the use of fire-stop materials must be followed.

Code Table 1 is based on common conditions of proper cabling and alignment of conductors where the length of the pull and the number of bends are within reasonable limits. It should be recognized that for certain conditions, a larger size conduit or a lesser conduit fill should be considered.

There shall not be more than the equivalent of four quarter bends (360° total) between pull points (boxes, conduit bodies, etc.)

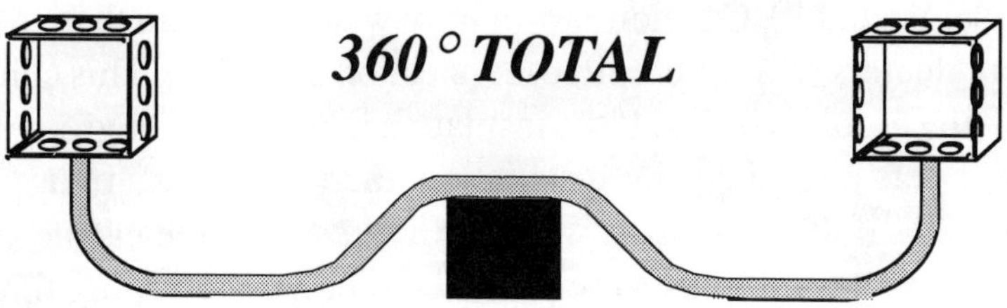

Penetrations of conduits through fire resistant walls and shaft enclosures, and floor, ceiling, and roof elements of fire resistant assemblies shall be protected in accordance with the Code.

The definition of **THROUGH-PENETRATION FIRESTOP SYSTEM** is a system installed to prevent, for a prescribed time period, the passage of flame, heat and hot gases through openings, which penetrate an entire fire assembly in order to accommodate cables, cable trays, conduits, etc.

Conduits of any material not harmful to concrete within the limitations of the code shall be permitted to be *embedded in concrete* with **APPROVAL OF THE ENGINEER**, provided they are not considered to replace structurally the displaced concrete, except as provided in the code.

Conduits of **ALUMINUM** shall NOT be embedded in structural concrete *unless* effectively coated or covered to prevent aluminum/concrete reaction or electrolytic action between aluminum and steel.

Conduits passing through a *slab, wall, or beam* shall NOT impair significantly the strength of the construction.

Conduits, with their fittings, **EMBEDDED WITHIN A COLUMN** shall not displace more than **4%** of the area of cross section on which strength is calculated or which is for fire protection.

Except when plans for conduits are approved by the engineer, **CONDUITS EMBEDDED** within a slab, wall, or beam (other than those merely passing through) shall satisfy the following:

1. They shall not be larger in outside dimension than 1/3 the overall thickness of slab, wall, or beam in which they are embedded.

2. They shall not be spaced closer than three diameters or widths on center.

3. They shall not impair significantly the strength of the construction.

Conduits shall be permitted to be considered as replacing structurally in compression the displaced concrete provided:

1. They are not exposed to rusting or other deterioration.

2. They are of uncoated or galvanized iron or steel not thinner than standard Schedule 40 steel pipe.

3. They have a nominal inside diameter not over 2" and are spaced not less than three diameters on centers.

CHAPTER 2

EXAM

QUESTIONS ON DRILLING & NOTCHING WOOD FRAMING MEMBERS

1. A good rule to follow is that a notch have a length no greater than ____ of the member's depth.

(a) 1/3 (b) 1/4 (c) 25% (c) 1/6

2. In exterior load bearing wall a wood stud may be notched to a depth not exceeding ____ of its width.

(a) 40% (b) 25% (c) 60% (d) 30%

3. A hole not greater than ____ of the stud width may be drilled in any wood stud.

(a) 40% (b) 25% (c) 60% (d) 30%

4. For a 2 x 8 floor joist, the maximum notch depth between supports and maximum notch length shall not exceed ____ and ____.

(a) 1 1/8" and 2 3/8" (b) 1 5/8" and 1 1/2" (c) 1 1/4" and 1 7/8" (d) 1 5/8" and 1 1/2"

5. A notch in a 2 x 10 floor joist shall not exceed ____ the depth of the member.

(a) 1/2 (b) 1/3 (c) 1/6 (d) 1/4

6. Notches are cut in both the compression and tension sides of the 2x4.

OKAY ☐ VIOLATION ☐

7. Shown below, the notch is cut in the tension side.

OKAY ☐ VIOLATION ☐

8. A 2x6 joist measures 5 1/4" in depth, a 2" diameter hole in drilled into the joist.

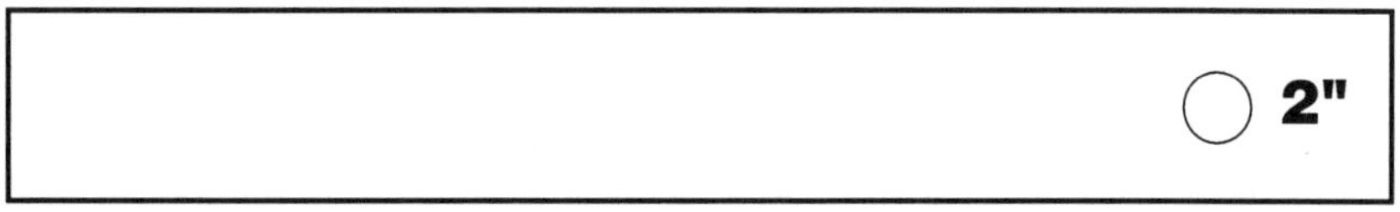

OKAY ☐ VIOLATION ☐

9. A 2x4 stud in a load bearing wall has a notch cut 1" deep.

OKAY ☐ VIOLATION ☐

10. A 2x8 floor joist has two 3/4" diameter holes drilled as shown below.

OKAY ☐ VIOLATION ☐

QUESTIONS ON BOXES AND RACEWAYS - EXAM 2

•Using your Code book circle the correct answer and write the Code section where found.

1. Where a permanent barrier is installed in a pull box, each section is considered as _____.

(a) permanent barriers are not allowed **(b) a separate box**
(c) 60% of the box **(d) the same box**

2. In straight pulls, the length of the box shall be not less than ____ times the trade diameter of the largest raceway.

(a) 4 (b) 6 (c) 8 (d) 12

3. What size rigid PVC conduit schedule 40 is required for eight #6 XHHW conductors?

(a) 3/4" (b) 1" (c) 1 1/4" (d) 1 1/2"

4. The volume per #14 conductor required in a box is ____ cubic inches.

(a) 2.25 (b) 2 (c) 3 (d) 2.5

5. In completed installations each outlet box shall have a _____.

(a) receptacle (b) switch (c) cover (d) fixture

6. Plaster, drywall or plasterboard surfaces that are broken or incomplete shall be repaired so there will be no gaps or open spaces greater than ____ inch at the edge of the fitting or box.

(a) 1/16 (b) 1/8 (c) 3/16 (d) 1/4

7. Enclosures supported by suspended ceiling systems shall be fastened to the framing member by mechanical means such as _____.

I. clips identified for use II. screws III. rivets IV. bolts

(a) I only (b) II only (c) II and IV only (d) I, II, III and IV

8. Conduit bodies shall have a cross-sectional area at least ____ that of the largest conduit to which they are connected, #6 conductors and smaller.

(a) 100% (b) twice (c) 40% (d) 75%

9. Where NM cable (romex) is used, the cable assembly, including the sheath, shall extend into the box no less than ____.

(a) 1/2"   (b) 3/4"   (c) 1/4"   (d) 1"

10. ____ boxes shall not be used where conduits or connectors requiring the use of locknuts or bushings are to be connected to the side of the box.

(a) Round   (b) Shallow   (c) Device   (d) Gang

11. The internal depth of outlet boxes intended to enclose flush devices shall be at least ____.

(a) 1/2"   (b) 7/8"   (c) 15/16"   (d) 1 1/2"

12. In combustible walls or ceilings, the front edge of an outlet box or fitting may set back of the finished surface ____.

(a) 1/4"   (b) 1/8"   (c) 1/2"   (d) not at all

13. A 3" x 2" x 2" device box is how many cubic inches?

(a) 12   (b) 14   (c) 10   (d) 8

14. Angle pull dimensional requirements apply to junction boxes only when the size of the conductor is equal to or larger than ____.

(a) #0   (b) #4   (c) #3/0   (d) #6

15. The number of #12 conductors permitted in a 3" x 2" x 1 1/2" deep device box is ____.

(a) 6   (b) 5   (c) 4   (d) 3

16. Metal plugs or plates used with nonmetallic boxes shall be recessed ____.

(a) 3/8"   (b) 1/2"   (c) 1/4"   (d) 1/8"

CHAPTER 3

THE SERVICE

T H E S E R V I C E

INSPECTING THE SERVICE

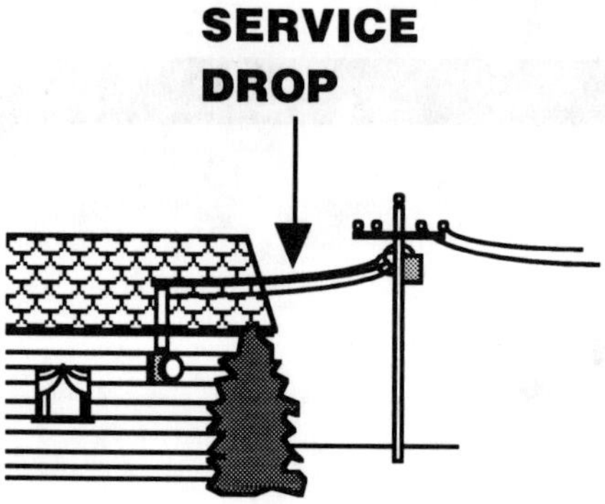

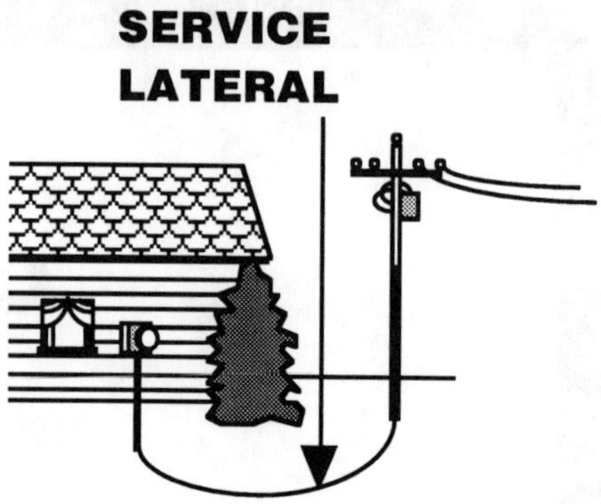

Service conductors supplying a building or other structure shall not pass through the inside of another building or other structure. This is a very important rule of safety that must be followed.

The overcurrent protection for service conductors is on the primary side of the transformer. In the event of a fault condition, the fault would have to burn clear in most cases. This is the reason that service conductors have separate rules from branch circuits and feeder conductors.

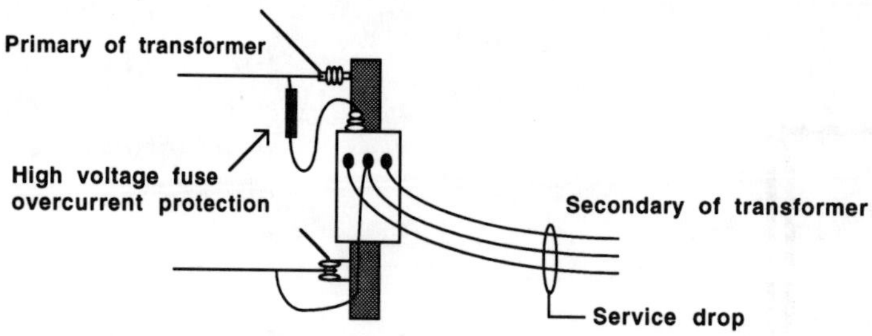

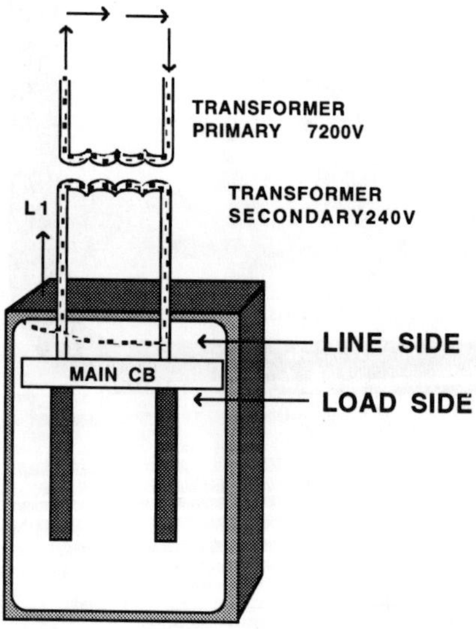

• **IF A GROUND-FAULT or SHORT-CIRCUIT occurs on the LINE side of the MAIN CB it does NOT see this current. The protection in these cases is provided on the primary side (high-voltage)of the transformer**

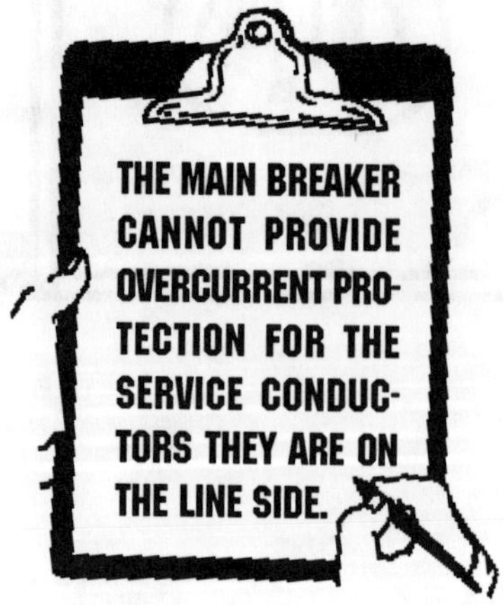

Where a service mast is used for the support of the service drop conductors, it shall be of adequate strength, such as *rigid conduit*, to support the conductors in the worst of conditions with the weight of ice and snow. In somes cases a guy wire is required for extra support.

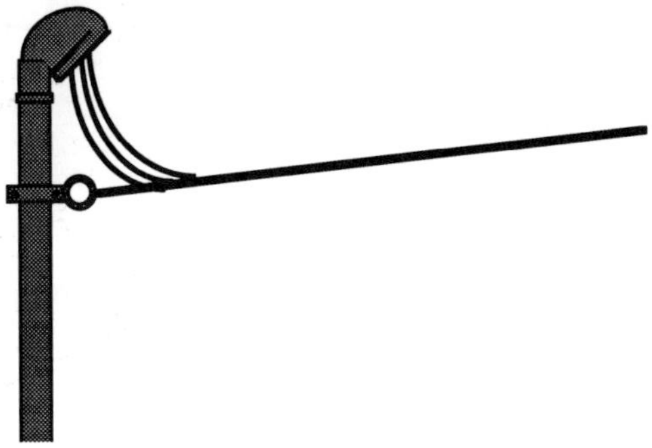

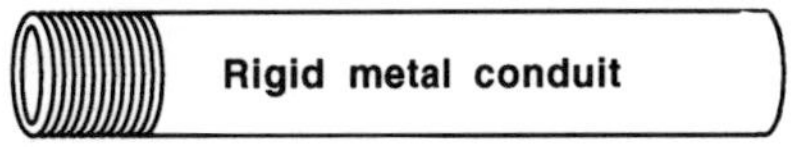

The Code does NOT allow other than service conductors being attached to a service mast. A separate mast or different vertical support is needed for cable TV, telephone, etc.

Raceways enclosing service entrance conductors shall be raintight and arranged to drain. The definition of raintight is: So constructed or protected that exposure to a beating rain will not result in the *entrance* of water.

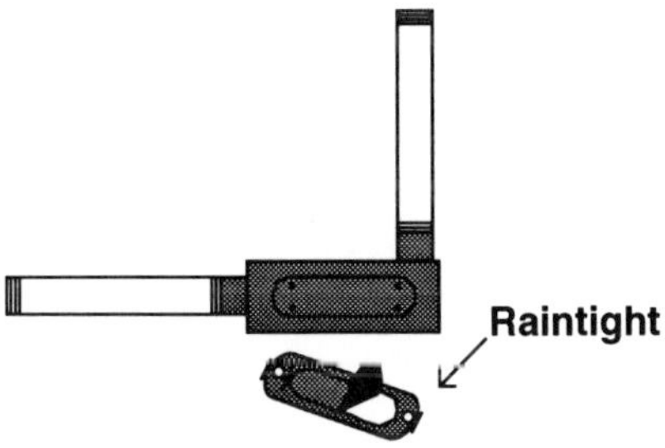

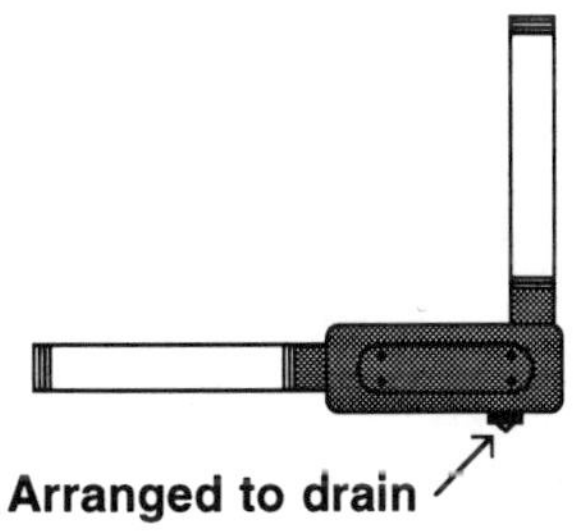

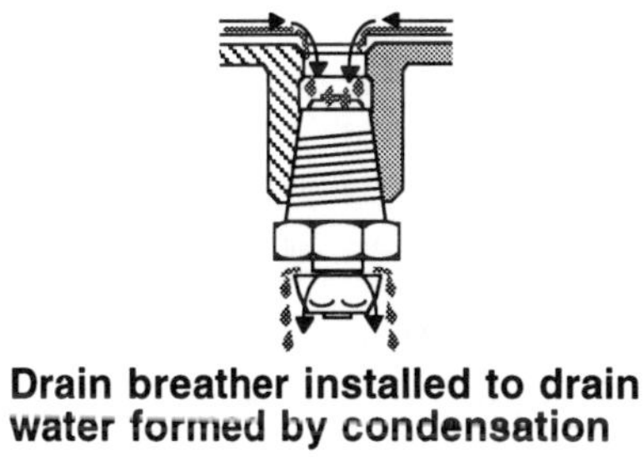

Another way to provide a weep hole is to drill a small hole in the bottom of a fitting. Although the Code requires that all unused openings in boxes and conduit bodies to be *effectively closed*, it is referring to knockout holes. Drilling a small hole to drain the fitting would be the same as the unused hole that can be used for mounting a box. But, this violates the listing and should be approved by the inspector.

Service entrance *cables* shall be supported within 12" of the service head, gooseneck or connection to a raceway and at intervals not exceeding 30".

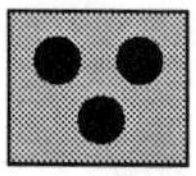

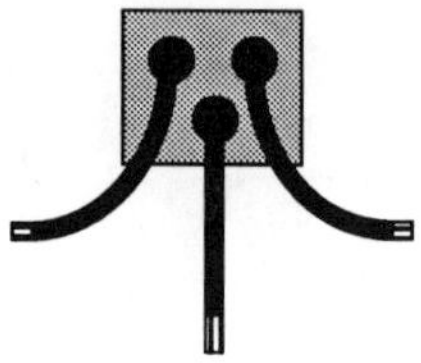

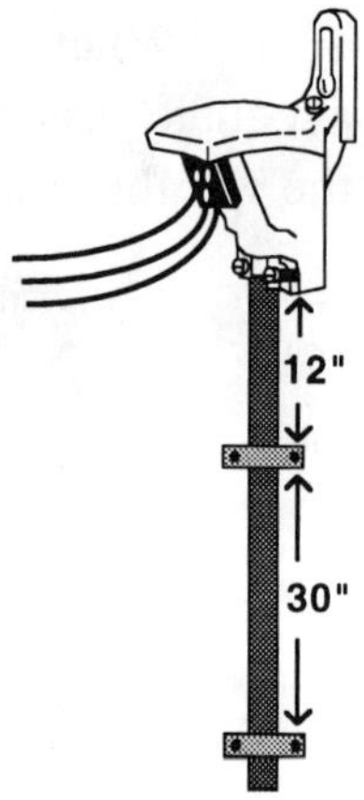

Each service conductor shall be brought through a separate bushed hole in the service head.

A *raintight* service head shall be installed to service raceways at the point of connection to the service drop. Service entrance cables shall have a *raintight* service head or sealed gooseneck.

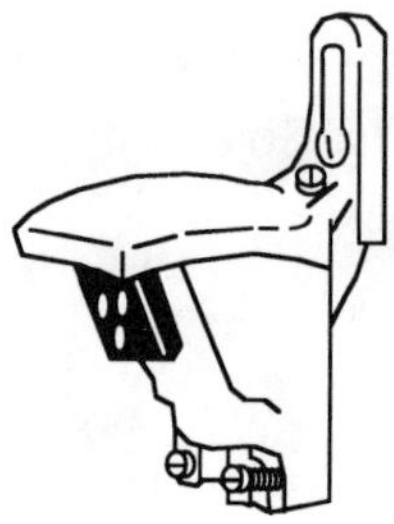

Definitions in the Code contain several "*proof*" and "*tight*" terms such as: Dustproof, Dusttight, Rainproof, Raintight,Weatherproof, Watertight, etc.

The easy way to remember the difference between "proof" and "tight" is **PROOF** is to prevent from **INTERFERING** with successful operation and **TIGHT** prevents the **ENTRANCE**.

Not all fittings that seem to be watertight are listed for watertight applications. Rubber and neoprene gland-type fittings are NOT listed as raintight due to variations in the outside diameters of the service entrance cables. Not all compression fittings for conduit are listed raintight.

Service heads and goosenecks are to be located above the point of attachment to the service drop. When it is impracticable to put the service head above the point of attachment it is permitted to be not more than 24" from the point of attachment.

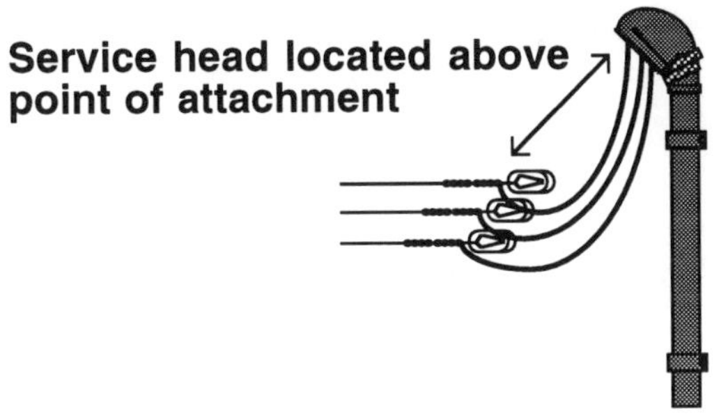

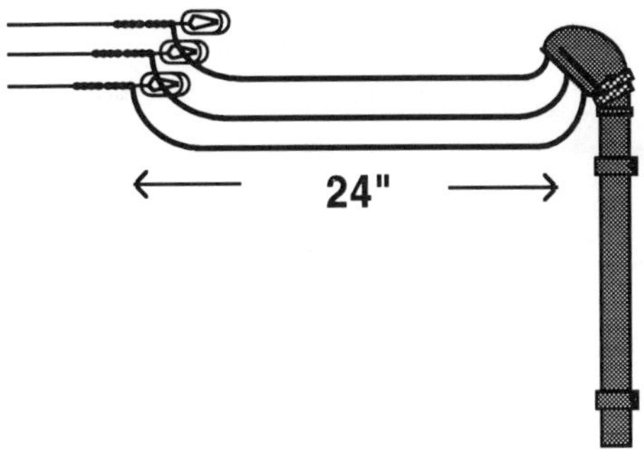

Service entrance conductors are required to be connected to the service drop conductors below the level of the service head or below the termination to the service cable. Drip loops are required to prevent water from following the conductors into the service head.

Service conductors shall be installed to keep water from entering the service raceway or equipment. The service head should be installed plum and not at an angle allowing rain to enter the enclosure.

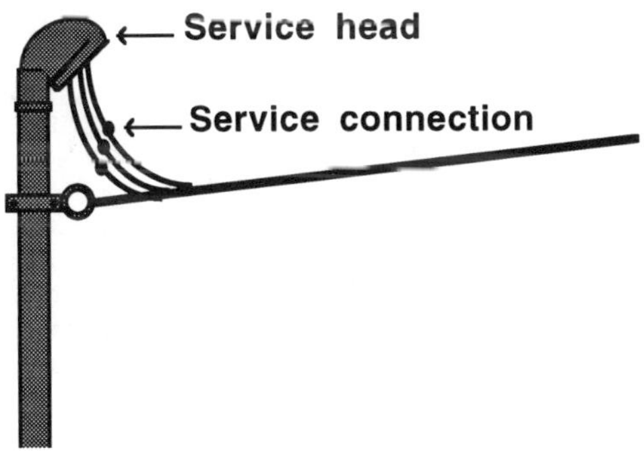

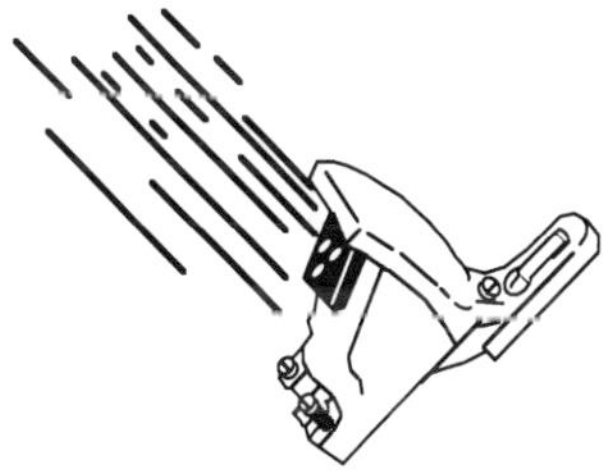

Service cables and open conductors shall be attached to the building by identified fittings for use with service conductors. The point of attachment of the service drop conductors shall not be less than 10'.

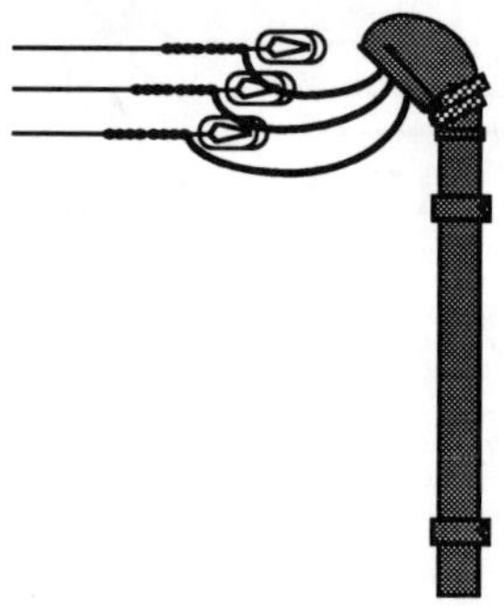

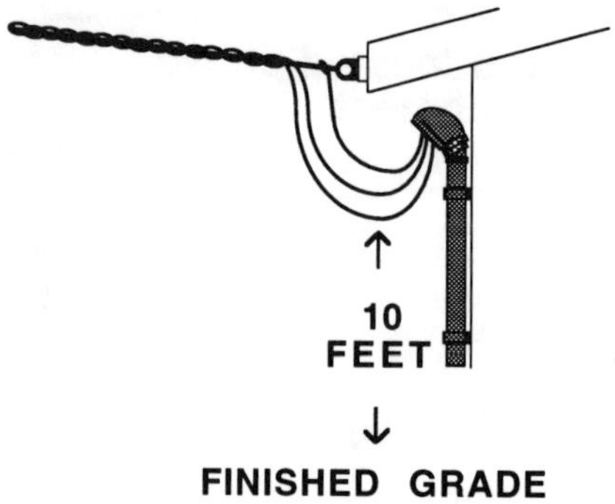

Service drop conductors shall have a vertical clearance of not less than 8 feet above the surface of the roof.

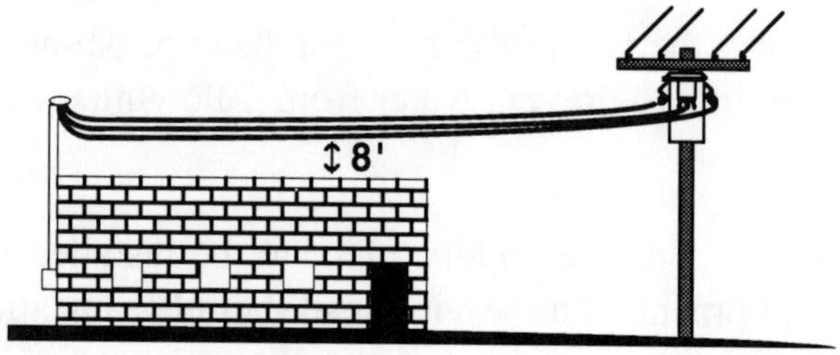

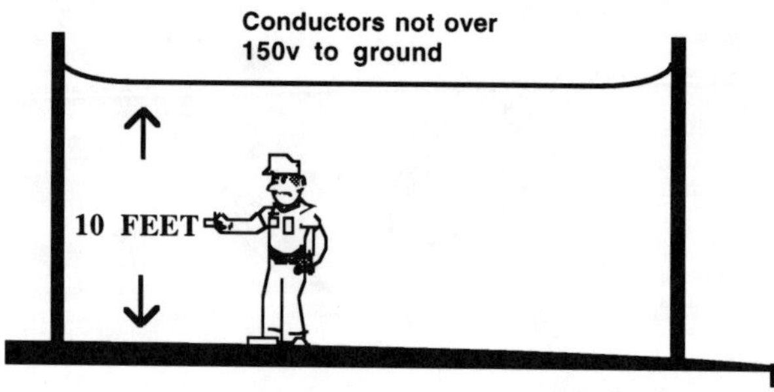

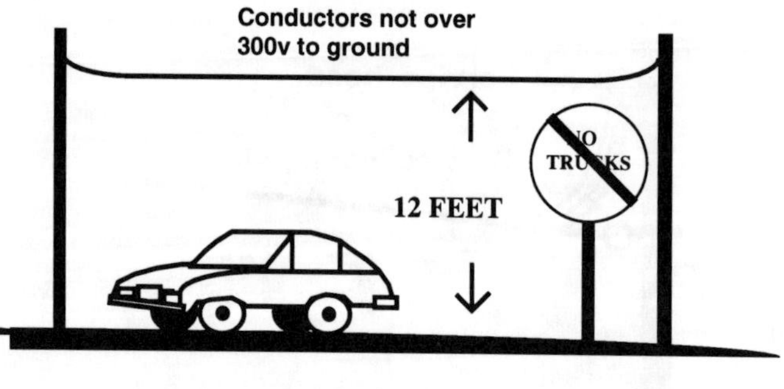

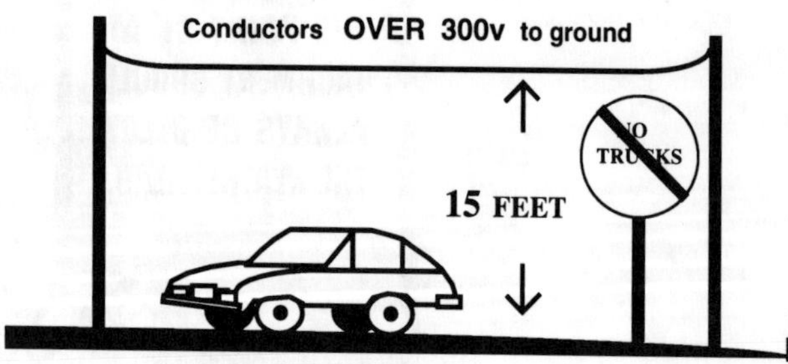

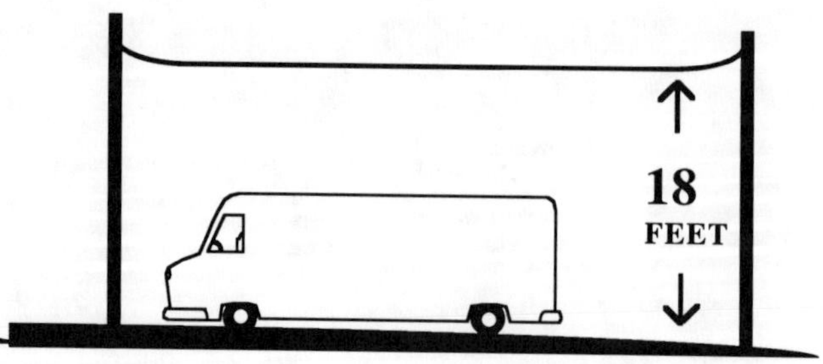

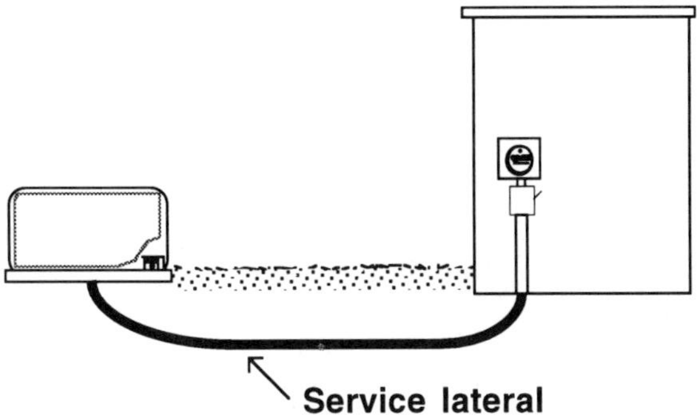

Service lateral conductors shall follow the burial depths of Table 300-5.

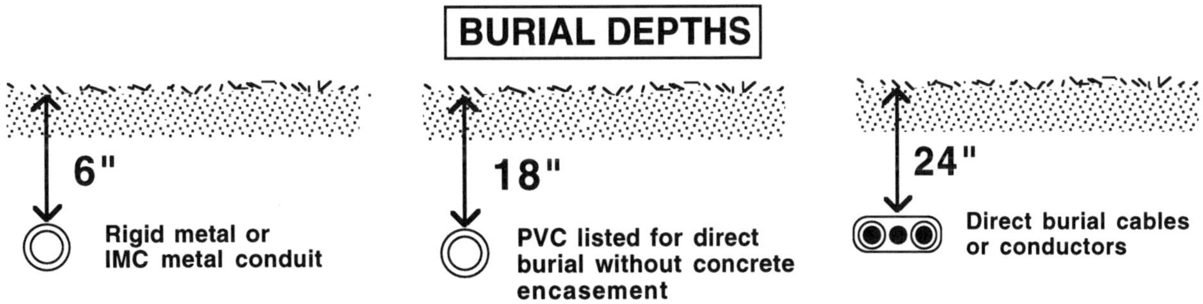

ENCASED

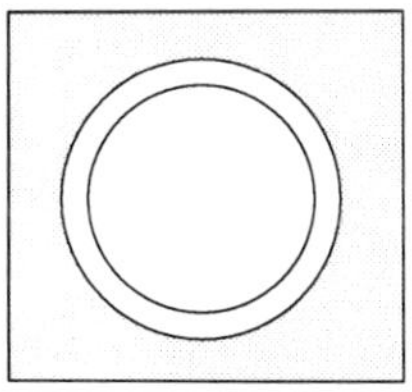

ENCASED means it must have concrete or mortar *in contact* on *all sides* of the raceway.

The Code in some cases refers to a trench below 2" thick concrete or under minimum of 4" thick concrete. Burial depths can be reduced for concrete *placed over* cables or conduits.

PLACED OVER

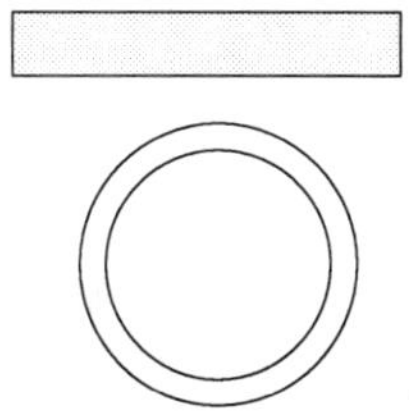

The concrete placed over a cable or conduit is to alert someone digging that the concrete has a protective function. Installed cable or conduit under a *sidewalk* does NOT comply with the Code rule. Someone digging under a sidewalk would not realize the sidewalk had a protective function.

Raceways embedded in concrete shall be arranged to drain.

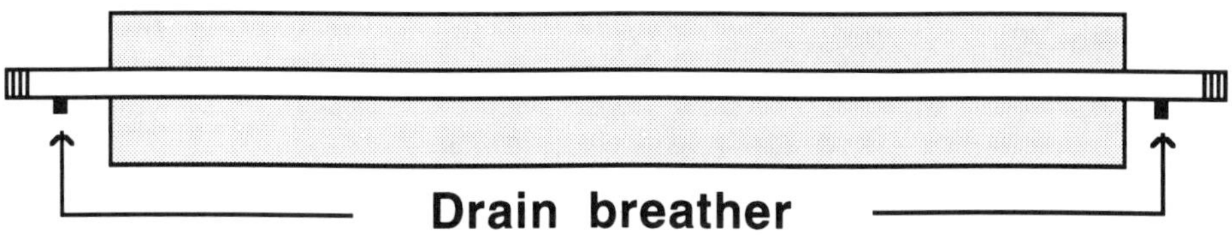

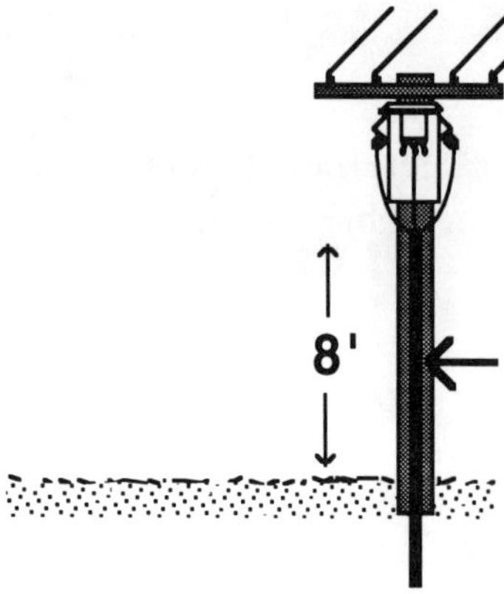

Conductors emerging from the ground shall be protected up to 8' above grade by rigid metal conduit, IMC or rigid schedule 80 PVC or the equivalent.

Service entrance above ground are required to be protected against physical damage. Where installed near sidewalks, walkways, or other locations where they could receive damage, they must be installed in a raceway of rigid strength.

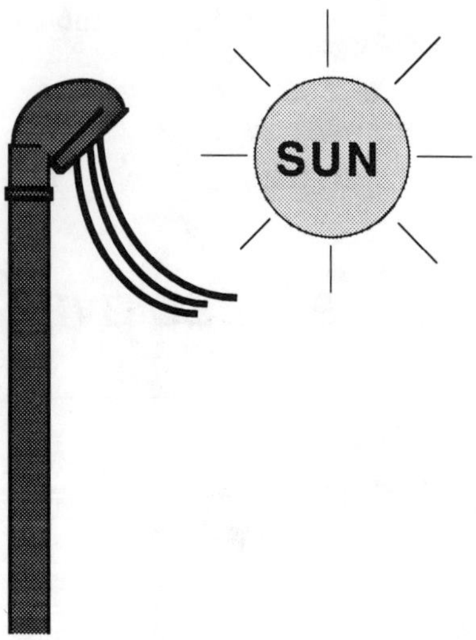

The service conductors shall be sized with an ampacity to carry the load without an increase in temperature rise. They shall also have adequate mechanical strength.

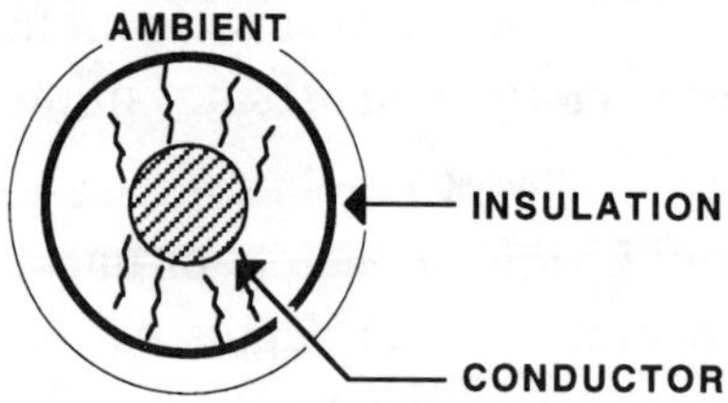

Service conductor ampacity should be corrected by a 115°F ambient *minimum*.

On very hot days and exposure to the heat is why circuit breakers trip in some cases. If the conductors were initially *sized correctly* to the **HEAT** encountered this nuisance tripping problem could be resolved.

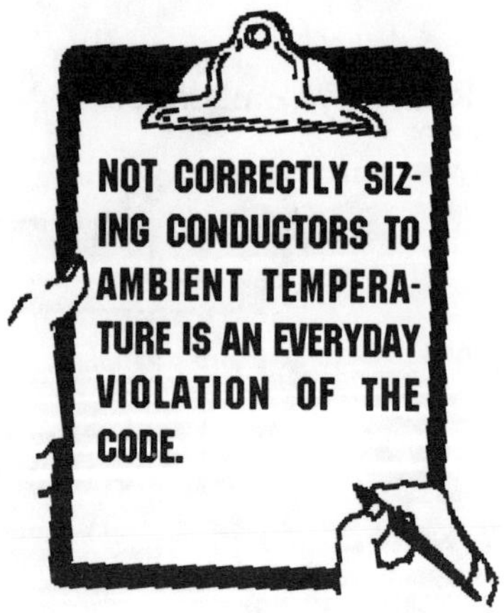

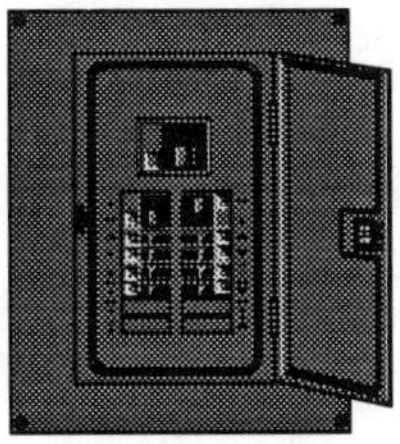

The Code requires electrical equipment to be installed in dedicated spaces, protected from damage, have easy access, and have working clearances.

A Code requirement that is often overlooked in residential construction is service equipment must be suitable for the **SHORT-CIRCUIT CURRENT AVAILABLE** at the terminals.

Service conductors have NO meaningful overcurrent protection so they must be protected as soon as they enter a building so they won't start a fire within the building.

If the service disconnect is located inside the building it must be located at the nearest point of entry to the inside of the building. There is NO maximum distance required in the Code. The distance is determined by the authority having jurisdiction.

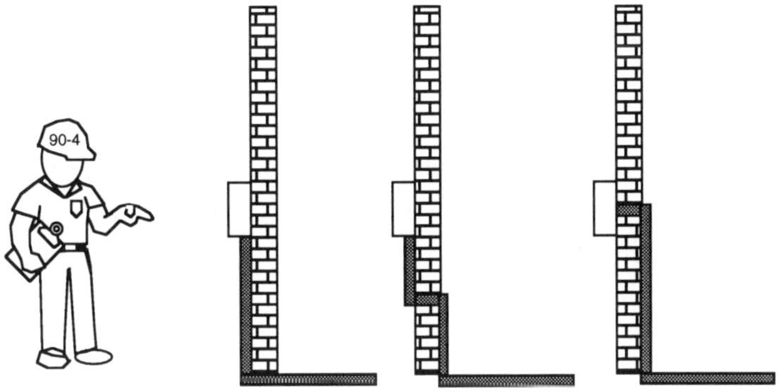

The general rule is the service overcurrent device shall protect all circuits and devices. There are six exceptions to the general rule.

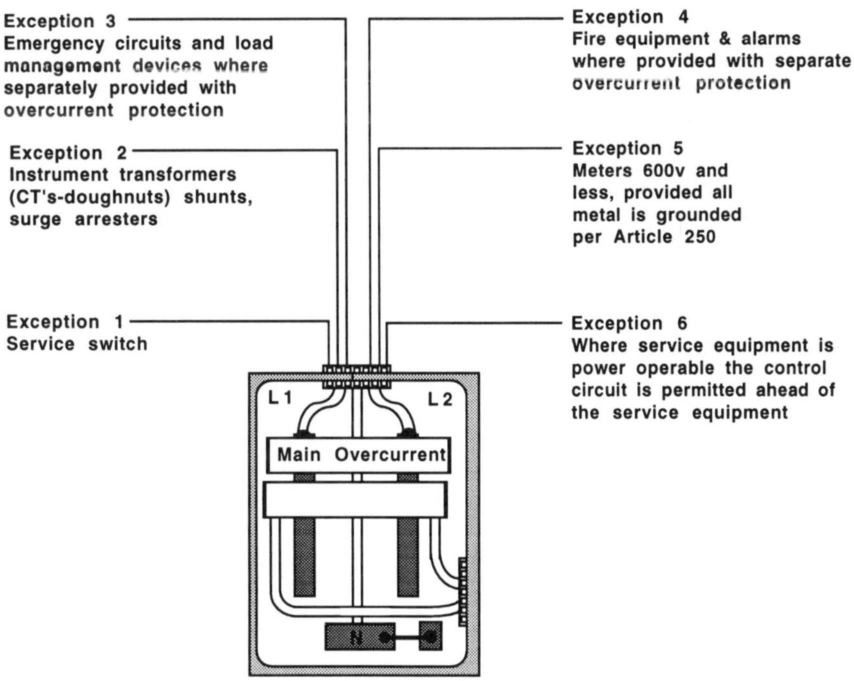

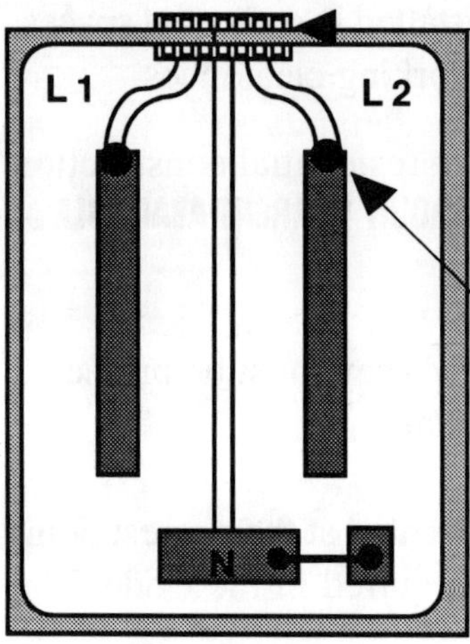

#4 AND LARGER CONDUCTORS MUST BE PROTECTED BY A INSULATED BUSHING OR THREADED HUB

TERMINALS LUGS ARE REQUIRED TO BE TORQUED PER 110-3b

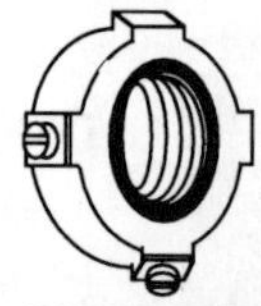

INSULATED THROAT BUSHING

TORQUE WRENCH

Overcurrent devices are NOT permitted in bathrooms or clothes closets of dwellings. Clothes closets have easily ignitible material and the bathroom may be occupied and the door locked.

Equipment that is not designed for outside use may become damaged by weather conditions while awaiting installation. This can produce a hazard when the equipment is placed in service.

Ground fault protection of equipment is required for a disconnecting means rated 1000 amperes or more on a 480/277v solidly grounded wye service.

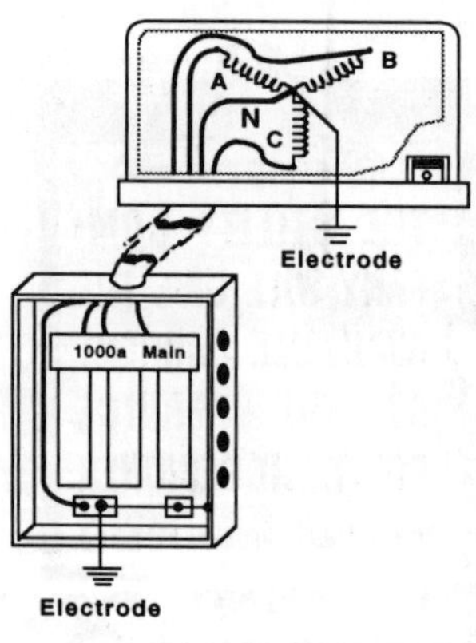

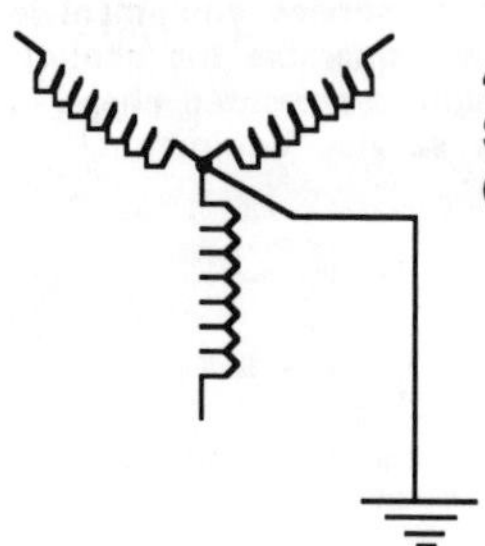

480/277v WYE SOLID GROUND

Overcurrent devices such as fuses or circuit breakers are sized to protect conductors to their ampacities. The overcurrent device will open when the current exceeds the rating. This opening of the overcurrent device can be caused by an overload, short-circuit or a ground fault.

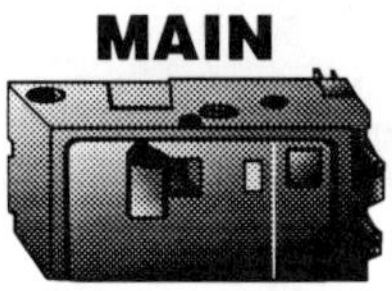

An inverse time circuit breaker operates on the principle that the *more* current that passes through it the *quicker* it trips, the lower the current the longer in time it takes to trip.

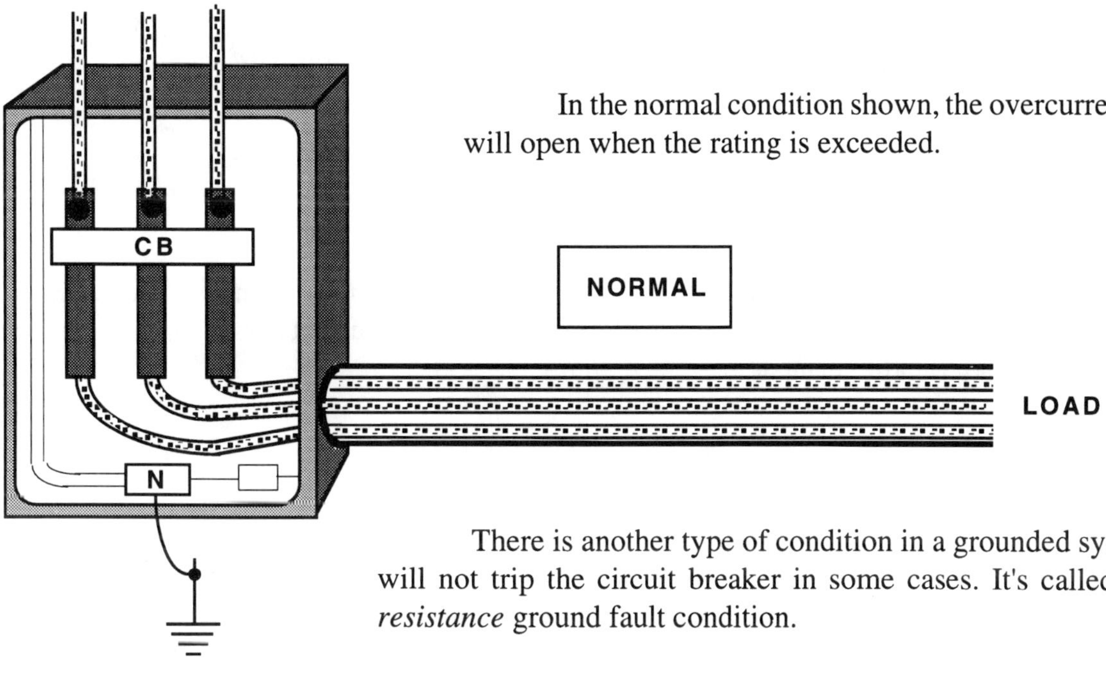

In the normal condition shown, the overcurrent device will open when the rating is exceeded.

There is another type of condition in a grounded system that will not trip the circuit breaker in some cases. It's called a *high-resistance* ground fault condition.

It's when a phase goes to ground (usually arcing) that current value is less than the rating of the overcurrent device so it never trips.

An example would be a feeder that has one phase going to ground on a metal conduit. If the overcurrent device has a rating of 1000 amps and the load current is 500 amps with a 300 amp fault current, the overcurrent device will not open until it reaches 1000 amps.

This is wasted energy which once it starts arcing, the intense heat can act like a torch causing major damage.

The Code requires a ground-sensor device which detects the ground fault current and will trip the overcurrent device preventing damage to the equipment. Arcing faults have been known to continue for over an hour at current levels too low to trip a breaker. The total energy released can lead to total destruction of the service equipment.

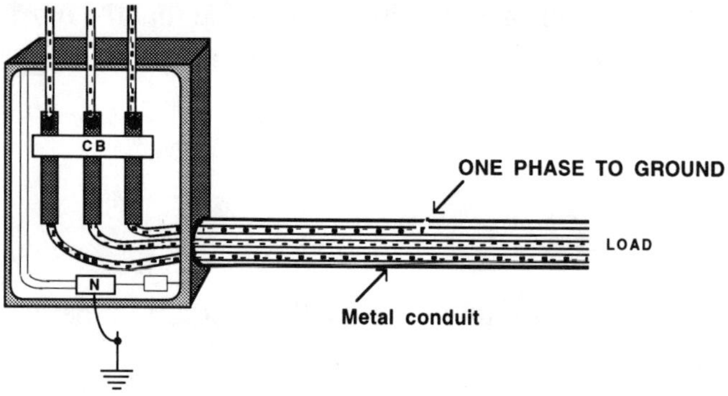

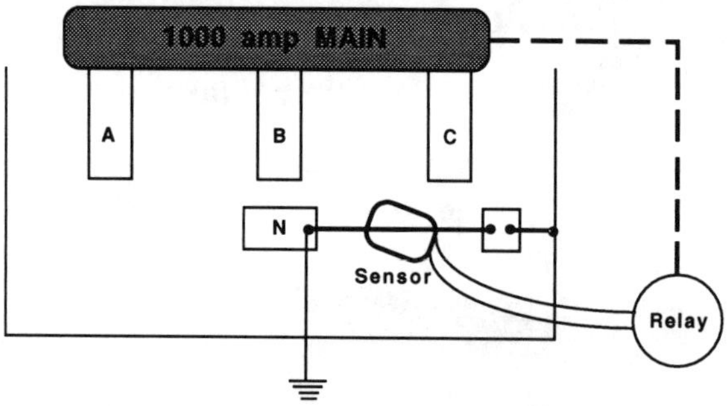

Ground faults occuring on a 208/120v wye are usually burnt clear and are not self-substaining. The peak line to neutral voltage is 120v x 1.41 = 170 volts, when the current passes through zero on the sine wave the arc is extinguished.

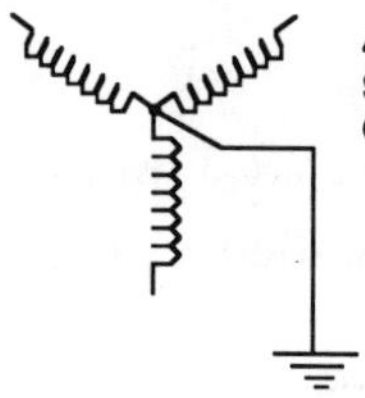

On a 480/277v wye system studies have shown that an *arcing* ground fault can be represented as a spark gap with a **SPARK-OVER VOLTAGE** of *375v*. 277v x 1.41 = 392 volts which is higher than the 375 spark-over level, so the arc will continue and will NOT extinguish itself.

Heavier loads and larger voltages occur at the service equipment. This is an area where damage and fire results in heavy losses. The inspector must do a thorough inspection of the service equipment. This is NOT an inspection to be taken lightly.

Faulty **CONNECTIONS** at the service terminals have been known to cause major destruction.

PROPER TORQUE

Because we live with the consequences of our work, we should pay especially close attention to even the smallest work detail.

The importance of properly torquing a connection cannot be stressed enough. Many fires have started due to the heat build up in an improperly terminated electrical conductor. The property owner is not knowledgeable in the electrical practices and depends on the electrician to install a safe electrical system.

When torque is applied to a threaded fastener, it produces a clamping force that holds the components together. Too much force, and the fastener will break or the material may be damaged. Not enough, and the assembly will not be held together properly. By controlling the amount of torque, the clamping or holding force is controlled.

All electrical connections must be properly torqued to prevent excessive heat build up caused by a loose connection or damage to the conductor caused by overtightening. **TOO LOOSE A CONNECTION OR TOO TIGHT A CONNECTION CAN CAUSE PROBLEMS.**

An electrical fire can start in a receptacle box that has nothing plugged into the receptacle. The receptacle is one of many receptacle outlets in the branch circuit. Loads may be plugged into other receptacles on this same circuit and current is flowing through this loose connection and heat continues to build up and causes arcing. At any splice, where a wire is connected by a screw or wire nut you have a possible electrical hazard. If you have a loose splice or connection, you have a high resistance connection which causes heat and can lead to arcing.

An electrical connection that is not tight enough will cause a high resistance to the current flow, which causes the connection to heat up. This extra heat causes even more resistance to the current flow.

An electrical connection that is **OVERTIGHTENED** can cause several problems. Overtightening can flatten the conductor to a point that it breaks. Overtightening the connection can also cause the connection to heat up. The conductor will push away from the stud area causing the conductor to cup which reduces the amount of contact with the lug. This smaller contact area causes overheating just as a loose connection will. Overtightening can damage the lug by causing small hairline cracks in the lug.

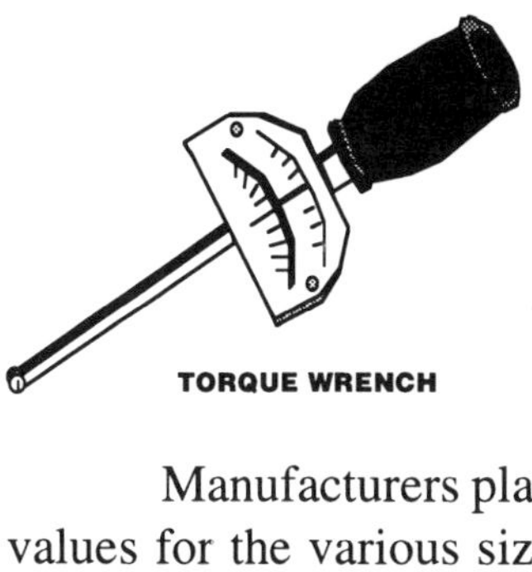

TORQUE WRENCH

Torque screwdrivers and wrenches are calibrated and marked in inch pounds or foot pounds and newton meters to show the torque or turning force. The amount of torque is adjusted on the screwdriver or wrench to the manufacturers recommended torque values. When the torque value is reached, a slipping will be felt or a clicking sound will be heard and the turning of the tool should be stopped at that point to prevent overtightening.

Manufacturers place a sticker inside electrical equipment to show the recommended torque values for the various size conductors. On some smaller equipment, the recommended torque is printed on the box containing the equipment. If the manufacturers recommendations are not available, the UL torque values available in UL publication number 486 should be used.

PROPER TORQUE

The size of the screw slot is important for proper torque. The chart below shows the U.L. recommended torque value for various screw slot and conductor sizes.

Circuit breaker terminals differ from switch and receptacle terminals in the tightening torque. You are required to torque to the *manufacturer's specifications*.

TIGHTENING TORQUE IN POUND INCHES FOR SLOTTED HEAD SCREWS #10 AND LARGER		
Conductor size	Slot width 0.047 or less and slot length 1/4 inches or less	Slot width over 0.047 inches or slot length over 1/4 inches
18-10	20	35
8	25	40
6-4	35	45
3	35	50
2	40	50
1	-	50
1/0-2/0	-	50
3/0-4/0	-	50
250-350 kcmil	-	50
400 kcmil	-	50
500 kcmil	-	50
600-750 kcmil	-	50
800-1000 kcmil	-	50

When installing electrical devices and equipment, some recommended torque values are given in inch pounds, some are given in foot pounds or in newton meters. Conversion of one value to another value is simple using the chart below.

MULTIPLY BY → / ↓ TO OBTAIN	INCH OUNCES	INCH POUNDS	FOOT POUNDS	NEWTON METERS
INCH OUNCES	1	16	192	141.6
INCH POUNDS	.0625	1	12	8.851
FOOT POUNDS	.00521	.0833	1	.7376
NEWTON POUNDS	.00706	.1130	1.356	1

FOOT POUNDS = $\frac{\textbf{INCH POUNDS}}{\textbf{12}}$ **INCH POUNDS = FOOT POUNDS X 12**

TORQUE SCREWDRIVER

PROPER TORQUE

Torquing of all electrical terminations is the fastest, surest way to eliminate connection problems. Use both torque scewdrivers and torque wrenches to tighten terminations to the prescribed contact pressure that assures the most reliable operation.

The UL (Underwriters Laboratory) Construction Materials Directory (green book) states the tightening torque. Some equipment may be marked to show a tightening torque for wire connectors intended for use with field wiring. Manufacturers are putting torque values in catalogs, specification sheets, and on boxes in which lugs and connectors are packaged.

A typical torque screwdriver has a torque range of 0 to 25 lb-in. The torque screwdriver has changeable tips for use on slotted screws, Phillips-head screws, hex-head bolts, and Allen-head bolts.

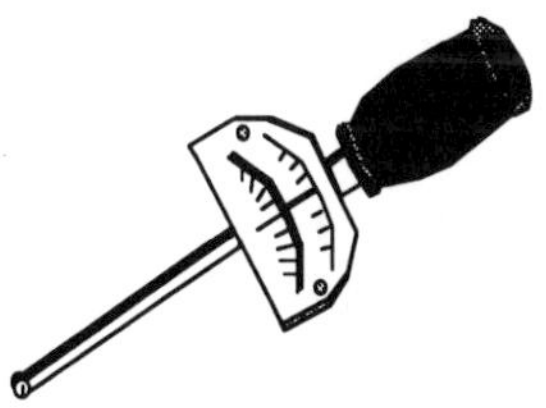

For torquing terminals used with conductors larger than #10, usual specified torque values exceed the 25 lb-in. capacity of the torque screwdriver, and a torque wrench must be used. There are two types of torque wrenches - the ratchet type and the beam type. A typical ratchet torque wrench has adjustable torque setting from 5 to 150 lb-in. capacity. A concentric sleeve on the handle provides for micrometer adjustment of desired torque.

The direct-reading type torque wrench is commonly used to tighten terminals for larger conductors, such as #500 kcmil to a torque of 300 lb-in. Because this wrench has a longer handle than the socket wrench, it can apply a given torque with less force required at the handle end than would be required for the shorter handle on the ratchet wrench. The beam wrench has a range from 0 to 600 lb-in. and can be used for torquing any terminations up to that value and for all terminations that require torque in excess of the 150 lb-in. capacity of the ratchet wrench.

Electrical connections properly made and torqued at the time of installation do not require periodic retightening. Once set, a termination should be left alone unless there is indication of malfunction, such as overheating.

The ideal torque value must be achieved as closely as possible to get the most effective connection, minimizing creep and cold flow of the conductor metal and prevent overheating due to increased contact resistance.

The electrical Code states in section 110-3b that listed equipment shall be installed in accordance with any instructions included in the listing. Inside the panelboard you'll find a label requiring the terminals to be torqued to a specified inch pound. Not to torque these terminals is a violation of the Code and the listing.

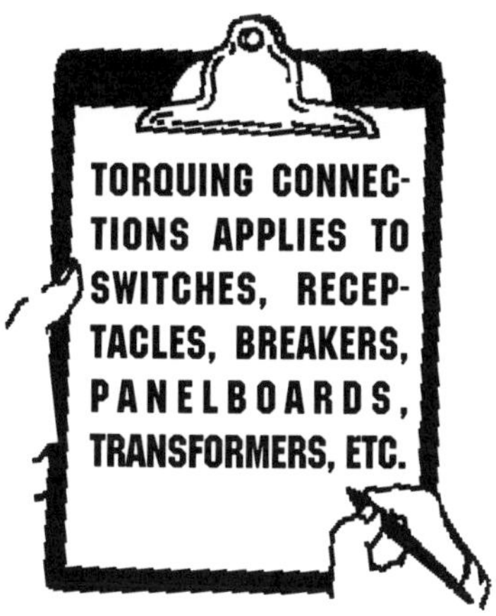

PROPER TORQUE

Over the years questions have been asked whether too much torque on a conductor termination would cause the same effects as a loose termination?

I would like for you to read the experts answers.

A loose conductor termination could cause arcing at the connection and result in a potential fire hazard. Too much torque on a conductor termination could deform the conductor or strip the threads of the terminal bolt or screw. If the bolt or screw threads were stripped, proper tension would not exist between the conductor and the terminal. Ultimately, due to expansion and contraction from heating and cooling of the terminal, the connection could become loose, and arcing could result. Thus, the result might be the same as with a loose connection.

The effects of too much torque are: First, the conductor itself or conductor strands will partially fracture and eventually split into two sections. One section will be retained within the termination device. The other section will become electrically isolated, or out of the circuit, and also become a physically loose strand that may flail about within the terminal area.

The resistance of each fractured strand is now increased due to the decreased cross-sectional area of that strand. Then, with no reduction of the total load current, the termination temperature will rise above its design limit. This thermal stress will cause any partially fractured strands to break apart completely, thereby increasing the thermal deterioration. The eventual result is the complete burnoff of the conductor from its termination, with the possibility of electric arcing and, as a very real consequence, high temperature plasma gas generation.

Second, on the clamping-type compression lug itself, the main effect will be from mechanical stress damage to the various parts of the termination due to this overtorquing. This unpredictably reduces the termination's current capacity. Because the load current is not reduced but is maintained, the temperature of the termination is raised beyond its thermal limits.

When the conductor termination is made tight to the point where the bolt is stressed into elongation and shear, causing a high-resistance point to develop. Over-tight terminals on insulation boards will cause the material to develop stress and cracks and eventual insulation breakdown. Over-twisted wire-nuts can cause problems. Always tighten to specifications.

The inspector should always spot check connections for proper torque.

Thermography and infrared surveys have also confirmed that torqing is essential in maintaining the proper current flow throughout the circuit.

In the early stages of a fault, the area of damage will show up as a hot spot with a small temperature rise. The more *advanced* the damage or degradation, the *higher* the temperature rise.

As a component continues to degrade, the temperature rises to the point where the developing fault now damages other components. For example, as a wire continues to heat up, more and more of its insulation and possibly the insulation of neighboring wires degrade. With enough degradation, the insulation will not prevent an arc between the wires possibly causing an electrical fire.

Intermittently operated equipment is likely to have the greatest damage. The reason is *thermal cycling*. When you turn equipment on and off, the components heat and cool, expand and contract. This has caused loose connections and arcing.

Improper torquing causes problems such as the unexplained breaker trips or the circuit has a low voltage.

Code section 310-4 refers to the paralleling of conductors and that they should be the same length, same size, same material, same insulation, and also they should be *terminated in the same manner.*

With the paralleling of conductors, even if they are cut to the same length, etc. if they are not torqued the same at the connection, heating can occur in one of the conductors. Due to improper torquing the problem accelerates as the load increases and causes phase overheating. It is critical that all three phases of a three-phase system be torqued properly to the recommended values.

When a wire termination is too loose or too tight to the point it causes a rise in current at that point, it is dissipating energy in the form of heat.

An actual case in an industrial plant proved that such a connection could cost $400 a year for each ampere it drew. With five connections you're talking $2000 a year in wasted electricity by heating an improperly torqued connection.

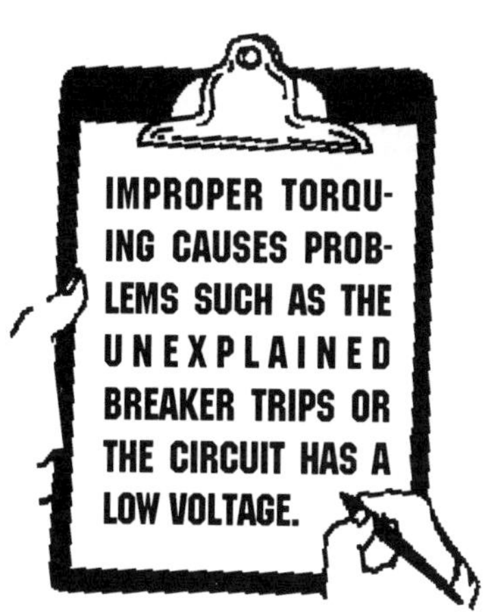

PROPER TORQUE

SMALLER SIZES OF ALUMINUM WIRE

Homes built before 1965 are unlikely to have aluminum branch circuit wiring. Homes built, rooms added, and circuits rewired or added between 1965 and 1973 may contain aluminum wiring.

Signs of trouble in aluminum wire systems include warm-to-the-touch face plates on the receptacles or switches, flickering lights, circuits that don't work, or the smell of burning plastic at the receptacles or switches. Unfortunately, not all aluminum wired connections provide such easily detected warning signs; aluminum wired connections have been reported to fail without any prior indications or problems.

When exposed to air, aluminum oxidizes. A thin, chemical insulation forms on the wire, increasing the resistance where it makes contact with the terminal. It has also been found that aluminum reacts chemically with brass terminal screws, causing corrosion. This increases contact resistance. Futhermore, temperature changes cause aluminum to expand and contract more than copper does. This causes connections to become loose and can increase contact resistance.

Aluminum wire is used today mainly in the **LARGER SIZE CONDUCTORS** used in the **SERVICE WIRING.** The connections are to be properly torqued and *maintained* as the years go by. The terminals must be marked "AL" or "ALUMINUM".

There has even been hazards reported using smaller copper wires. In several cases when #14 copper wires were spliced using steel crip sleeves, without twisting the conductors. The wires were forced apart in the crimping process, so the entire current was forced to pass through the steel sleeve.

In one case, the connection had overheated for so long that the connection had burned away almost completely, and one of the copper conductors had apparently vaporized for the first inch away from the connection, down to approximately a size #18 wire. The tape and insulation had been reduced to carbon powder.

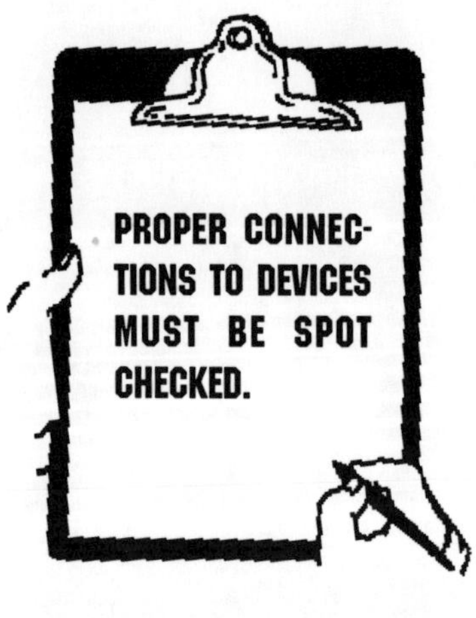

CHAPTER 3

EXAM

SERVICE INSPECTION QUESTIONS

1.

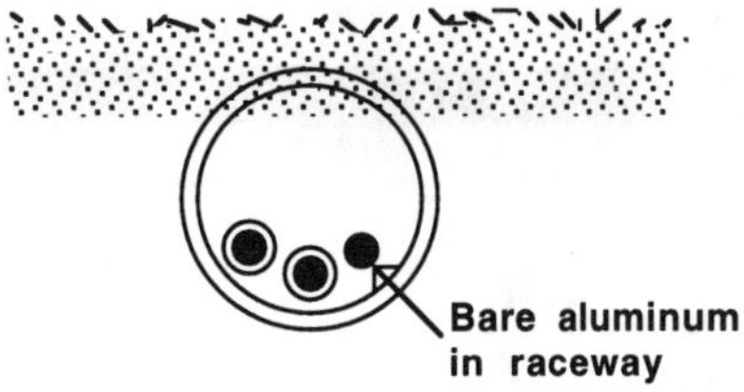

☐ OKAY ☐ VIOLATION

2. The grounding electrode conductor is installed in the same raceway as the service conductors.

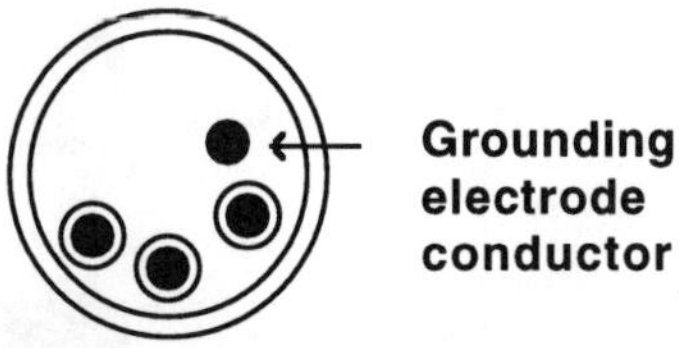

☐ OKAY ☐ VIOLATION

3. A single building has two services; a single-phase and a three-phase service.

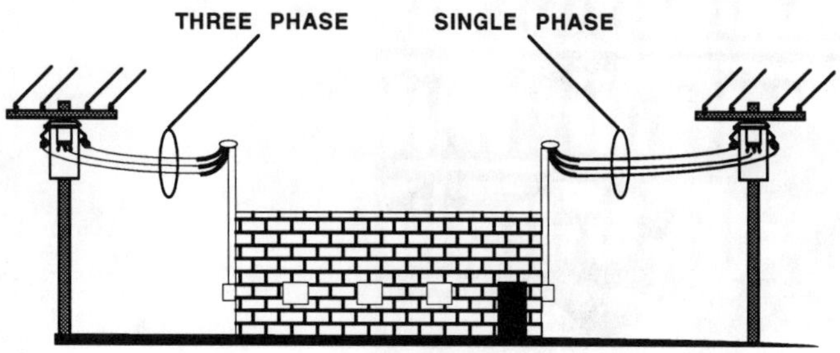

☐ OKAY ☐ VIOLATION

4.

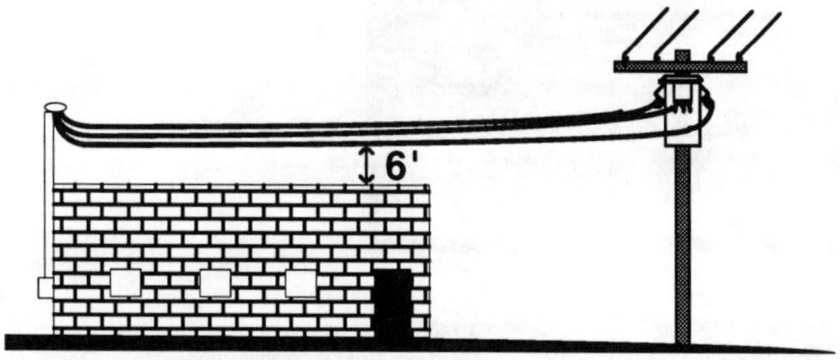

☐ OKAY ☐ VIOLATION

5.

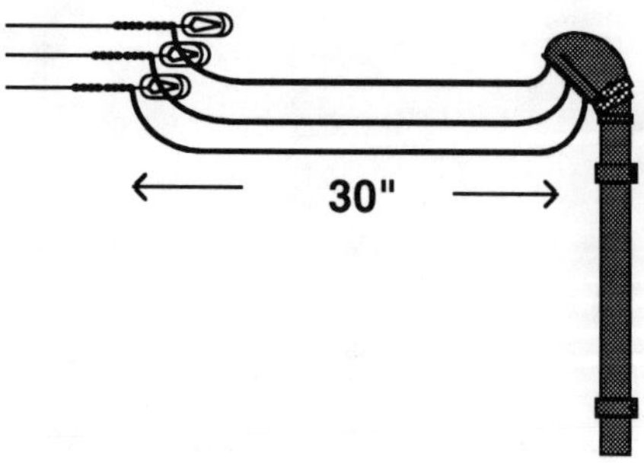

☐ OKAY ☐ VIOLATION

6. Electrical metallic tubing is used as a wiring method for service-entrance conductors.

☐ OKAY ☐ VIOLATION

7.

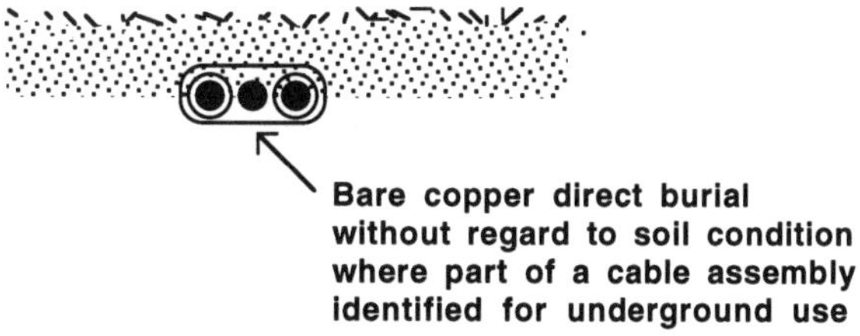

☐ OKAY ☐ VIOLATION

8. For a service supplying a single branch circuit, the disconnect shall have a rating not less than 30 amps.

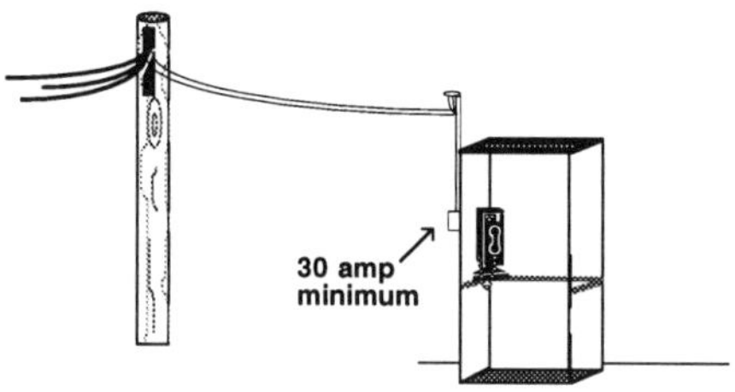

☐ TRUE ☐ FALSE

9. An overcurrent device protecting service conductors to a separate load, such as a water heater on a load management circuit, can be locked or sealed to prevent tampering.

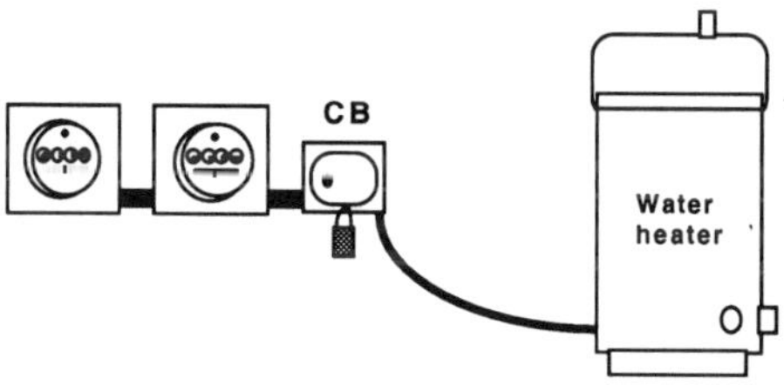

☐ TRUE ☐ FALSE

10. The Code allows eight movements of the hand to disconnect the service.

☐ TRUE ☐ FALSE

11. Service drop conductors open or multiconductor (triplex) shall have a clearance of not less than 6' from windows that can be opened, doors, porches, etc.

☐ TRUE ☐ FALSE

12. The grounded (neutral) conductor shall not be less than the grounding electrode conductor sized from Table 250-122.

☐ TRUE ☐ FALSE

13. If the service mast is located within 4' from the edge of the roof and the voltage between conductors does not exceed 300, a reduction in clearance to not less than 12" is permitted for the overhang.

☐ TRUE ☐ FALSE

14. No fuse or circuit breaker is permitted in the grounded conductor except a circuit breaker that opens all the service conductors simultaneously.

☐ TRUE ☐ FALSE

15. To make sure of the proper hookup of the ground fault protection system, a performance test is required in accordance with the instructions provided with the equipment. A written record of this test shall be provided for the inspector.

☐ TRUE ☐ FALSE

CHAPTER 4

GROUNDING

GROUNDING THE SERVICE

The sketch below shows the neutral bus in the service panel properly *bonded* to the panel and electrode. The service equipment enclosure is called the "bullseye" of grounding. The service equipment is the only location where all the grounding is connected together.

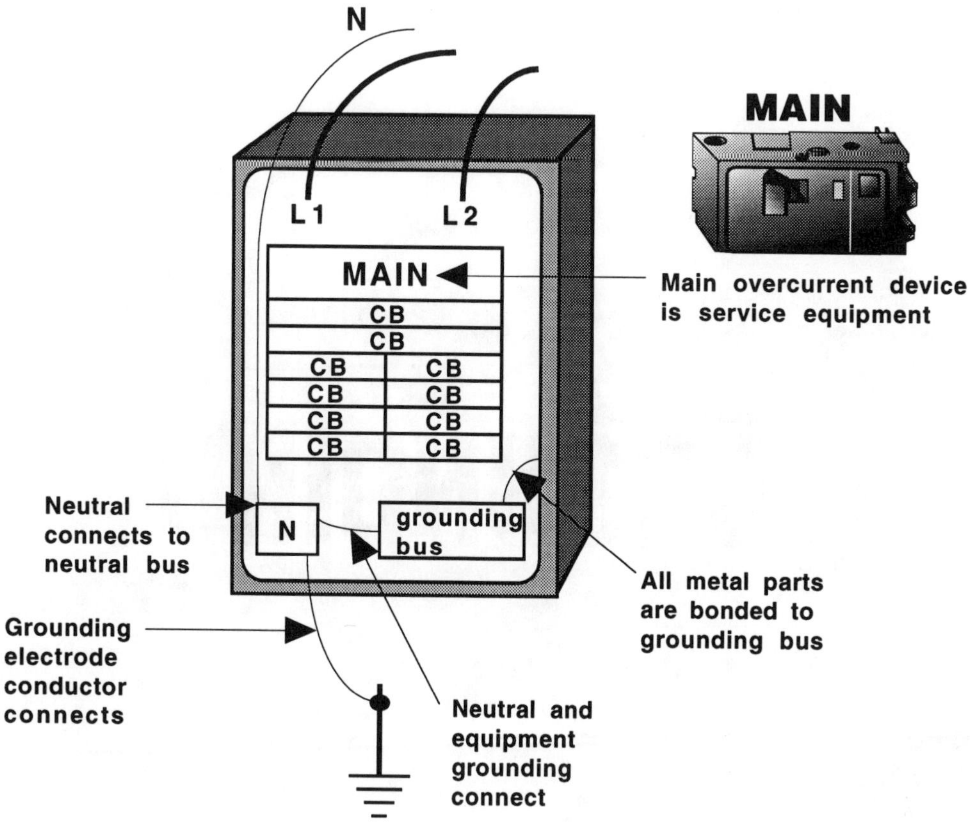

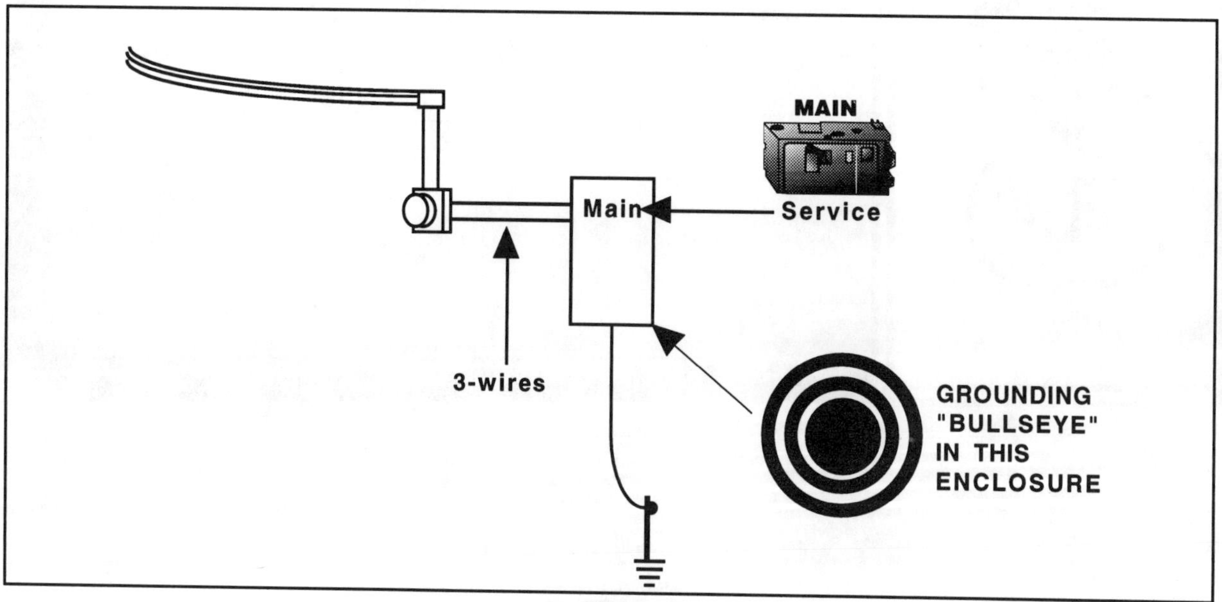

Code section 250-50 lists the *four* electrodes in parts a,b,c and d.

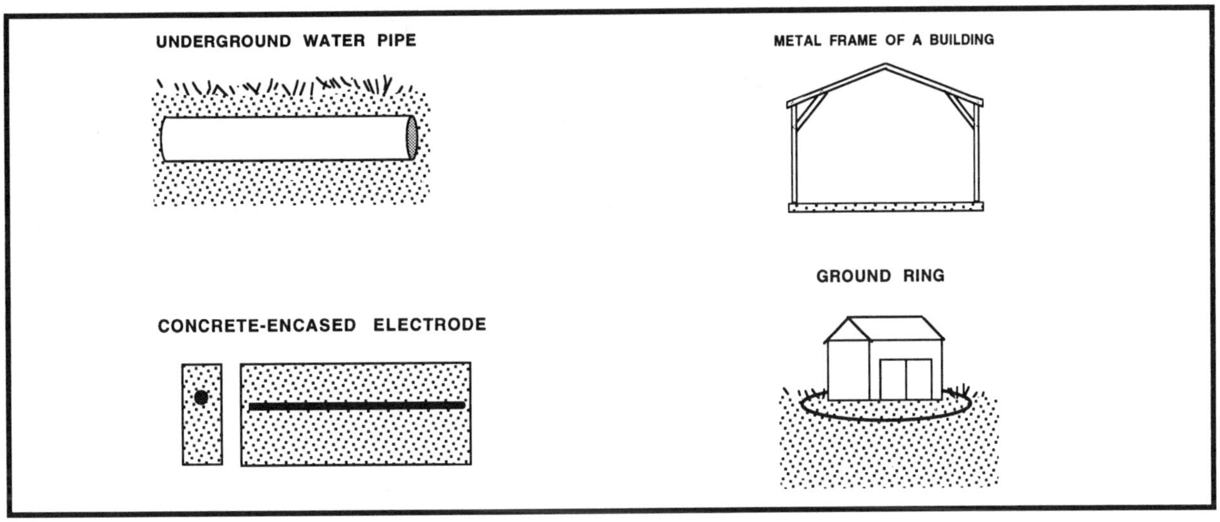

What I see in my travels is the 8' ground being driven to ground the service and the ground rod is NOT the best selection for an electrode. Only when *none* of these four are available should you consider driving a ground rod.

What is the best electrode for a low-resistance to earth ground reading?

Studies have shown that continuous underground metallic water piping systems have a resistance to earth of *less than three ohms.* The length of piping is an important factor, as the more piping you have in contact with earth the more area you have to discharge the electrons from a large surge. Local metal underground piping systems, metal well casings, metal building frames, and the like generally have a resistance to earth of substantially less than 25 ohms. For small systems where ground currents are relatively small the above electrodes are generally used because of the economical cost factor.

Buried plates are the *least* efficient type of the made electrode. Ground rods are generally used and are the most popular made electrode because of their cost. Plates are expensive compared to rods.

A ground rod driven in sandy type soil does not provide a tight connection (compaction) with earth. I have seen an 8 foot ground rod pulled from the sandy soil with the bare hand. A ground rod driven in sandy soil has a high earth resistance due to this poor contact.

Tests have proven that the **"UFER GROUND"** has very low-resistance to earth readings. Years ago in Arizona, which is usually hot and dry, and the soil being sand and gravel, 24 buildings were grounded with the Ufer ground. Resistance to earth readings were conducted every other month for 18 years. Records that were kept showed the maximum reading was 4.8 ohms and the minimum reading was 2.1 ohms. The average value for the 24 buildings over an 18 year period in the hot dry sand was 3.57 ohms! No servicing was required during this period. The installations used 1/2" steel rebar set in a concrete footing. As as result of these tests the Code accepted Mr. Ufer's suggestion that a *#4 or larger copper wire 20 feet long embedded in a concrete footing* be accepted as an electrode.

CONCRETE-ENCASED ELECTRODE

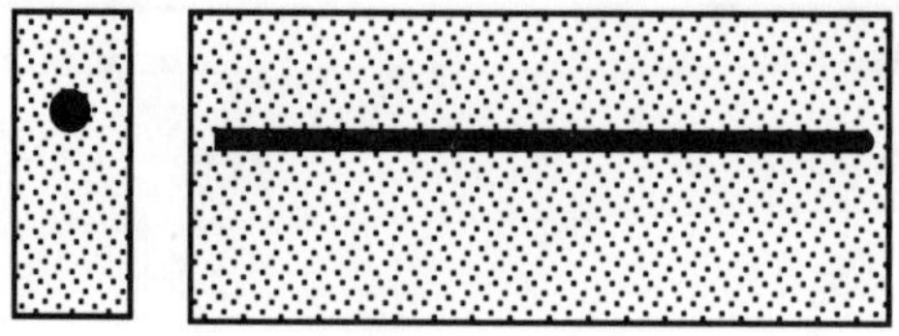

This is an excellent electrode for several reasons: (1) The rebar acts as multiple rods (2) The concrete is in contact with the earth over a large area. (3) The weight of the building keeps a good contact pressure to earth. (4) Concrete in contact with earth retains moisture. Do *not* use visqueen or any plastic barrier between the concrete and earth when pouring the footing. This would defeat the direct contact with earth. Tests have shown in some installations the resistance was less than *one ohm.*

We made earth tests on our new building here in Orlando, Florida in the sand and during the dry season. The lowest reading recorded was 24Ω which was connected to the rebar steel in the concrete footing. We read 39Ω connected to over 100 feet of 1" copper main water system. I had also installed in the concrete footing a bare #4 copper conductor 20 feet in length, it read 30Ω. We drove a 5/8" copper ground rod to a depth of 8 feet and it read 480Ω. We then drove a second rod 6 feet away and tied the two ground rods together and had an earth reading of 260Ω. These actual tests and readings are shown on video.

How can none of the four primary electrodes *not* be available? Well, let's say the building is wood frame and has no steel. The water piping is PVC and not metallic. And for some reason the concrete footing has no rebar steel, etc. Why didn't you install a ground ring?

Do you honestly think a driven ground rod is **EQUIVALENT** to the mass of concrete footing in contact with earth, which has the weight of the building on it. Which will meet the 25Ω or less Code requirement? You're the inspector.

A good local code rule is to require **BEFORE** pouring the concrete in the footing to inspect the rebar steel to assure the copper grounding electrode conductor has been connected to the rebar. This is the best ground, besides that you are following what the Code requires.

If the electrician is not on the job before the footings are poured, the question of the **AVAILABILITY** of a grounding electrode therein will be subject to the *interpretation* of the inspector.

As an electrician, I would rather connect to the rebar in the footing than to drive a ground rod.

In the past, the bonding connection between communication systems and the service was usually made by a connection to an interior metal water pipe. With the increasing use of PVC piping and sometimes the electrode being concealed, it's hard for the communication installer to find the service ground to bond his system. The Code requires that **all** systems must be bonded **together!** If a separate electrode is used to ground the telephone or cable TV, it must be connected to the service grounding electrode. **All** electrodes must be connected together to prevent dangerous voltages.

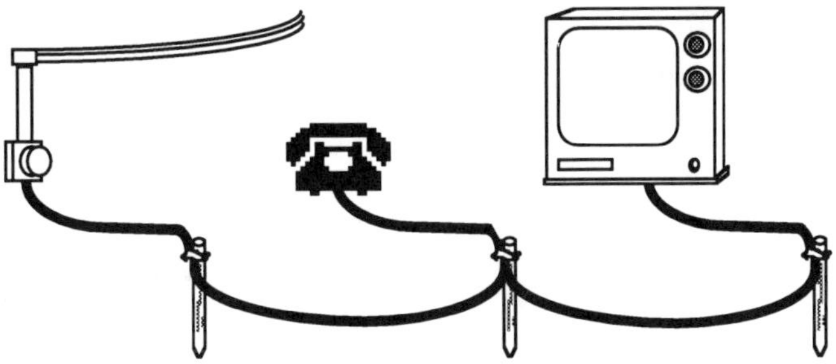

Many cable TV systems in houses are installed incorrectly. The cable installer unfamilar with the Code grounding rules connects the cable to the TV and drives a separate ground rod and connects the cable sheath to the ground rod. The receptacle suppying the 120v to the TV has a grounding wire (bare or green) and a grounded wire (white) of which both are connected to the service grounding electrode.

30,000 volts

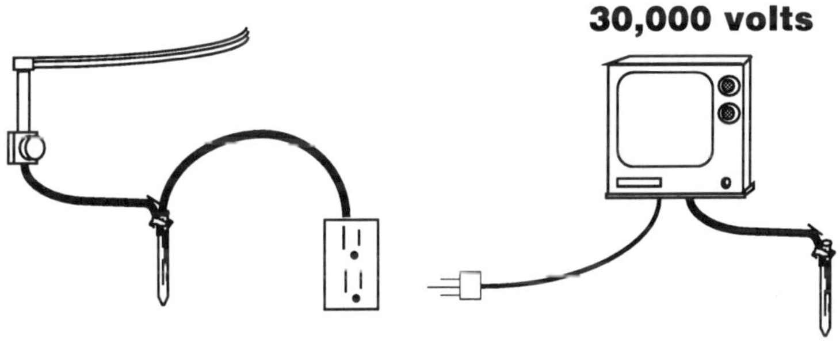

If the TV had a metal cabinet and a 3-wire cord plugged into the receptacle, large differences in voltage could occur between the sheath on the cable and the exposed metal on the TV.

If the service ground was 25Ω to ground and a lightning strike of 1200 amperes came through the grounded conductor through the earth resistance of 25Ω, the increase in voltage would be: $E = I \times R = 1200 \text{ amps} \times 25 \text{ ohms} = 30{,}000 \text{ volts}$. If the cable TV sheath had only a separate ground connection, the 30,000 volts could appear across any two separately grounded surfaces of the TV set. This is why during a lightning strike many TV's, telephones, computers, etc. get wiped out. There can only be **one** service ground and everything must tie to the bull's eye.

The solution is the cable sheath must connect directly to the equipment grounding conductor and **NOT** to a seperate electrode.

It is important to understand that a fuse or circuit breaker **cannot** protect against the low level ground faults that are killing people through electrical shock and are starting the number one cause of fires, **arcing**.

The grounding circuit is like a chain, any broken link and the safety is lost. As the years go by usage and abuse can cause an open or weak link in the chain.

Bonding minimizes differences in voltage between exposed metallic parts, and the equipment grounding conductor is to make a ground fault into a high level ground fault that will trip the breaker or blow the fuse.

WHY DO WE GROUND A CIRCUIT?

In the early days they learned that although the grounding of a circuit increased the fire hazard, it also decreased the life hazard.

The example below shows an **UNGROUNDED** circuit to a motor has a voltage of 120 volts.

To touch either wire on an ungrounded circuit would be a shock hazard. An ungrounded circuit with a balanced phase-to-ground capacitance has a normal line to ground voltage between any wire and ground.

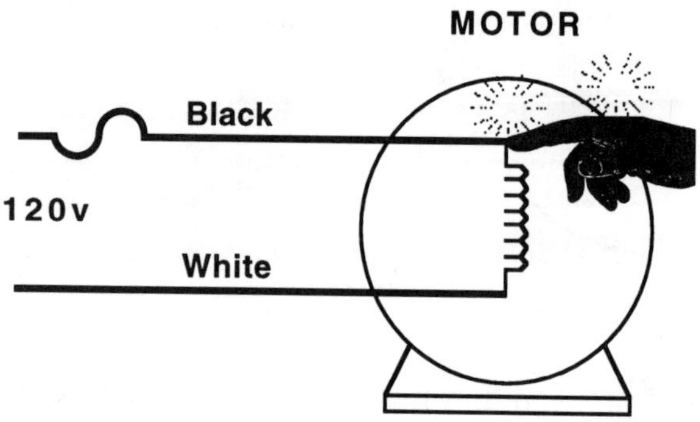

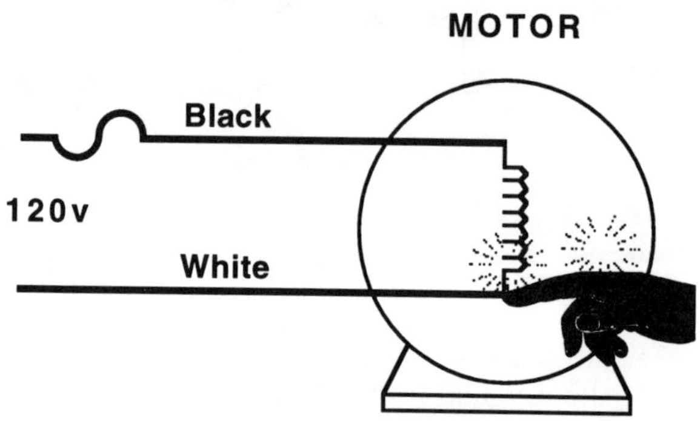

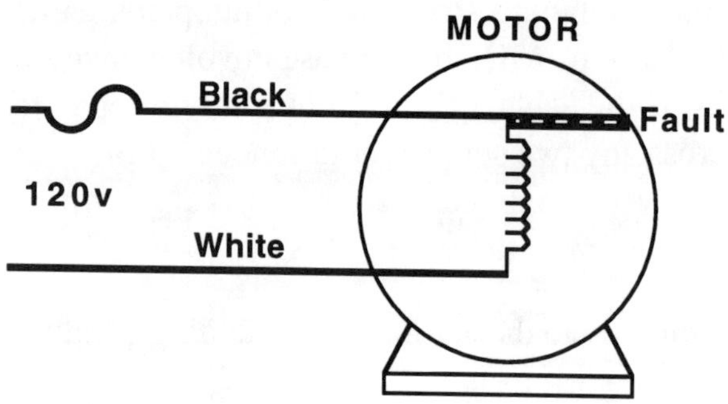

An insulation failure (fault to metal) on an ungrounded circuit does **NOT** generate enough current to open the overcurrent device, the phase-to-ground capacitance of the circuit will generate a small amount of current equal to the charging rate of the circuit. This amount of current is usually **NOT** enough to cause a spark or start a fire.

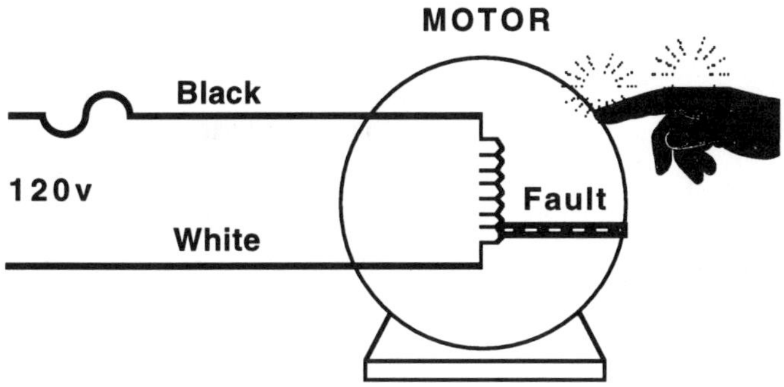

A person touching the metal motor frame would experience a shock.

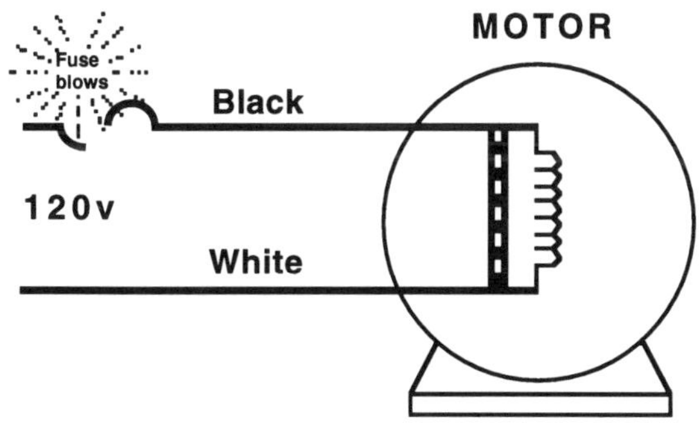

An insulation failure between the black and white wires (short circuit) would open the circuit.

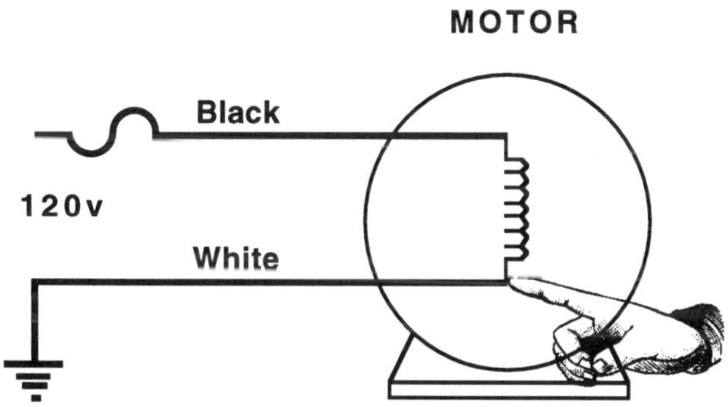

Now the white is intentionally **GROUNDED**. It takes earth's potential which is zero. A person standing on earth would have the same potential and should feel no shock hazard. **•IT'S NOT A SAFE PRACTICE TO TOUCH ANY WIRE YOU FEEL IS ENERGIZED.**

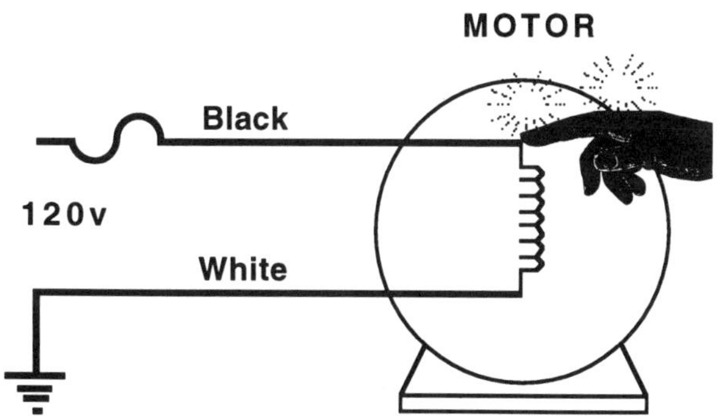

Touching the black (hot) wire you would experience a shock hazard. Now the person is the path to ground.

GROUNDING

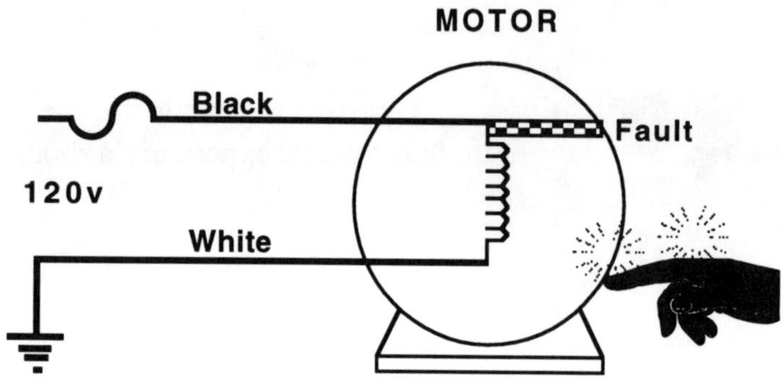

With an insulation failure in the black wire a person touching the metal motor frame would experience a shock.

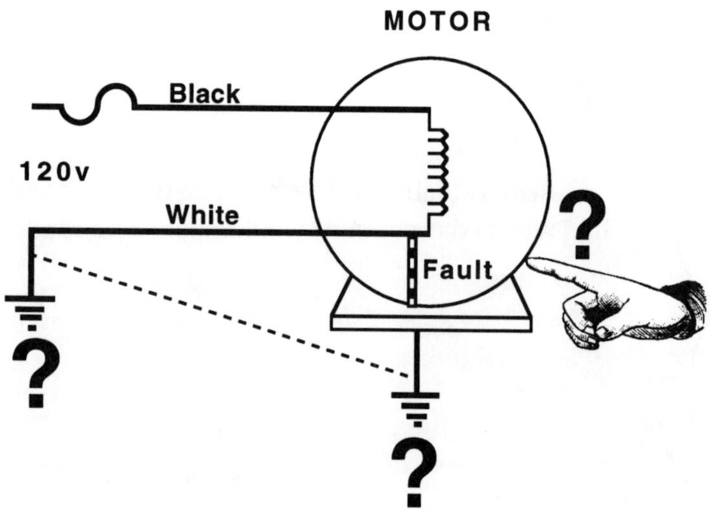

With an insulation failure in the white wire a person touching the metal motor frame experiencing a shock would depend on if the metal frame is at the same ground potential as that of the white wire. If there is NOT a good ground at either place you could experience a shock.

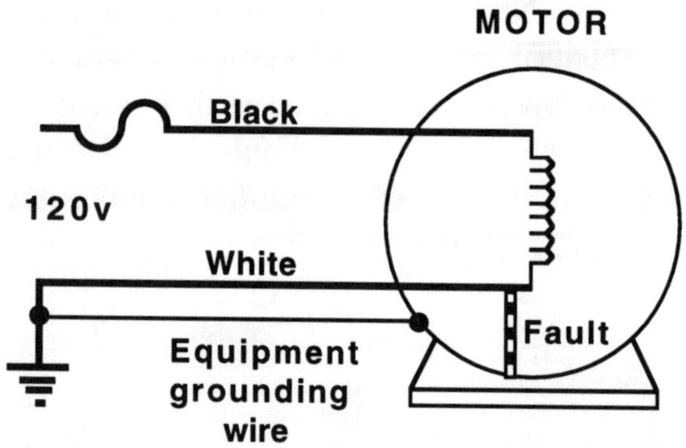

This is why an equipment grounding conductor is tied to metal parts and tied to the white wire at the service equipment. This keeps all equipment and metal parts at the same ground potential of zero if you have an **EFFECTIVE** grounding path.

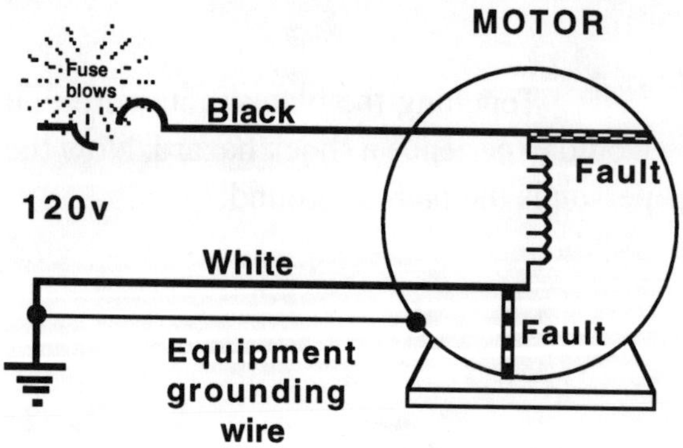

Now the fault current to the metal frame of the motor is carried by the equipment grounding conductor back to the panel to blow the fuse. Without the equipment grounding conductor the fuse would not blow with a fault current.

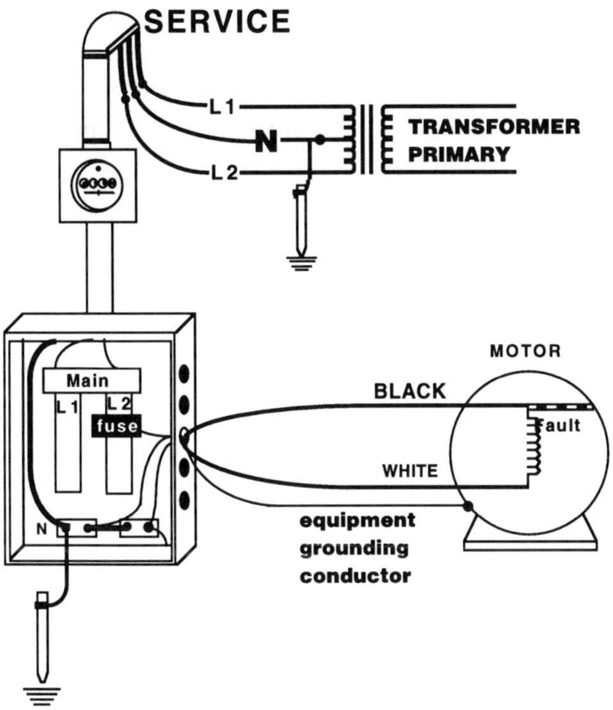

The fault current from the black wire to the metal motor frame will follow the equipment grounding wire back to the service panel to the neutral bus and follow the service neutral conductor back to the transformer and the fault current will return from the source through L2 conductor to the L2 panel bus bar and blow the fuse.

If any of the connections or conductors are loose or open in this effective grounding circuit it will effect the amount of current that reaches the fuse and will determine how quick it will blow.

We are to provide a path of the least impedance to allow heavy fault currents to follow this path and blow the fuse instantly.

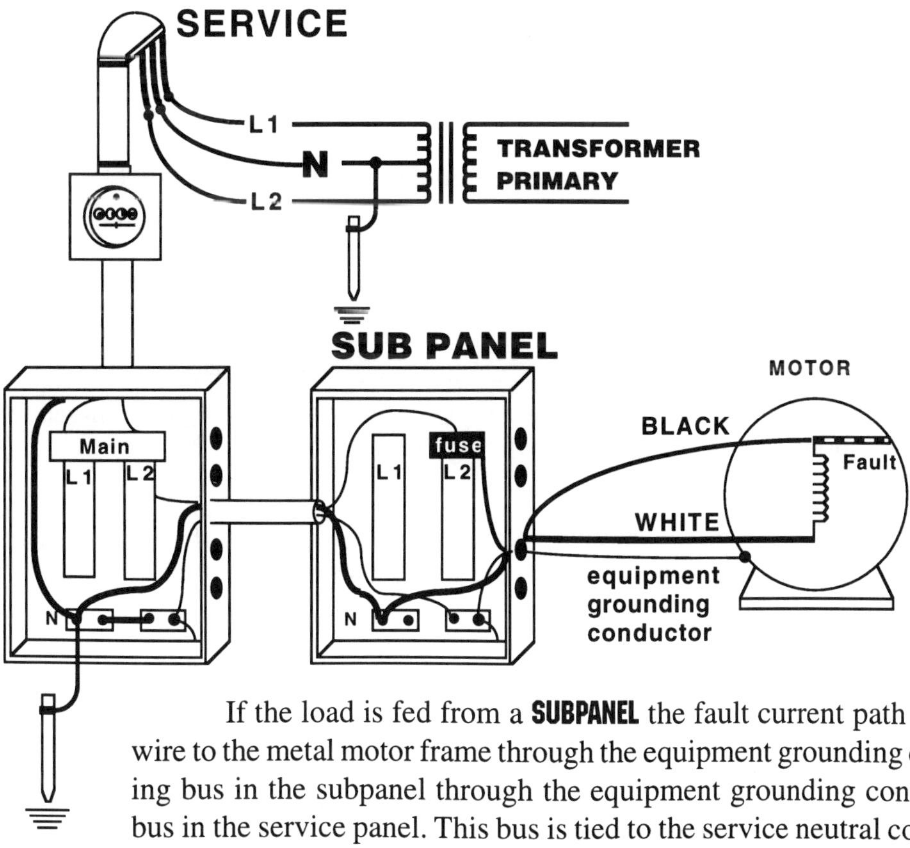

If the load is fed from a **SUBPANEL** the fault current path is from the black motor wire to the metal motor frame through the equipment grounding conductor to the grounding bus in the subpanel through the equipment grounding conductor to the grounding bus in the service panel. This bus is tied to the service neutral conductor which provides a low impedance path to the transformer and to the source back through L2 busbar in the service panel on to L2 bus in the subpanel to blow the fuse.

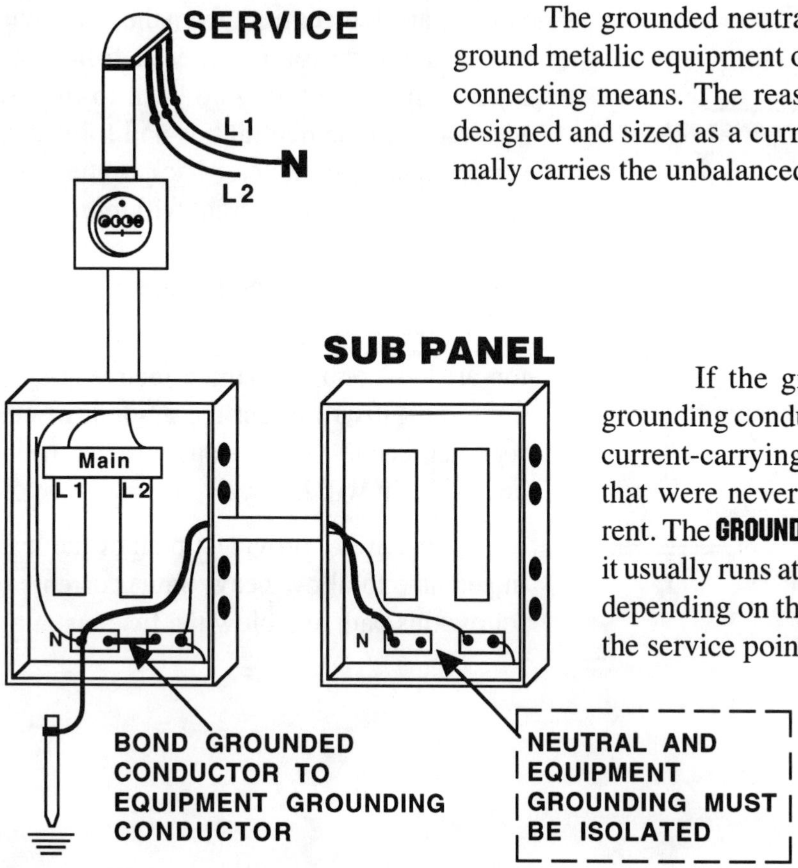

The grounded neutral conductor shall **NOT** be used to ground metallic equipment on the **LOAD SIDE** of the service disconnecting means. The reason for this is that the neutral is designed and sized as a current-carrying conductor, and normally carries the unbalanced current in the circuits.

If the grounded conductor touches the grounding conductor on the **LOAD** side it will make current-carrying conductors out of components that were never intended to routinely carry current. The **GROUNDED** conductor is **INSULATED** because it usually runs at some voltage level above ground depending on the amount of voltage drop back at the service point of grounding.

If the neutral were grounded downstream (subpanel) from the service disconnect it would be in **PARALLEL** with the equipment grounding conductors, and the neutral current would be shared by both the neutral and the equipment grounding conductors. This condition places a voltage on the frames of such equipment, the amount of which varies according to the voltage drop in the neutral circuit and the amount of unbalanced neutral current flowing.

The equipment grounding conductors (raceways, conduits, wires, etc.) are designed to carry fault current during a ground fault, but not for carrying current continuously. Therefore, this section is intended to prevent continuous current on the grounding system.

The neutral is grounded at the service equipment only. Any fault current that returns along the equipment grounding system can enter the neutral at the service, and the neutral provides a *low impedance* path for this fault current to return to the utility source. The Code permits the neutral to be used to ground equipment and enclosures on the **SUPPLY SIDE** of the service equipment only, because on the **SUPPLY SIDE** of the service equipment the neutral provides this *low impedance* return path to the source for a fault.

Both the grounding electrode conductors and the grounding connections must have an adequate cross section so that they can carry the **WORST** case of ground-fault current without overheating.

The low impedance path allows the high current flow when a ground fault occurs, causing rapid operation of the overcurrent protective device upstream from the fault.

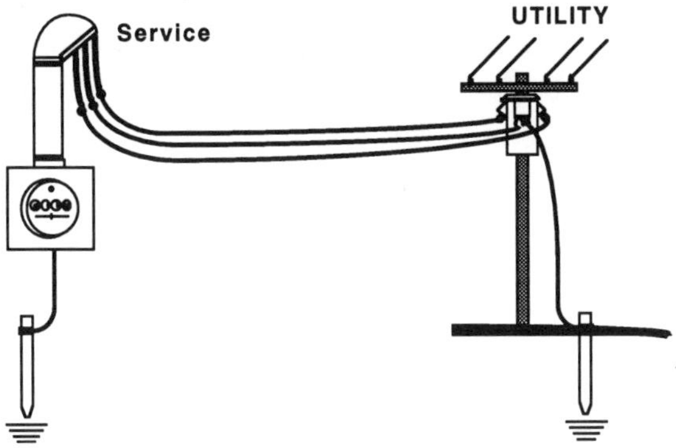

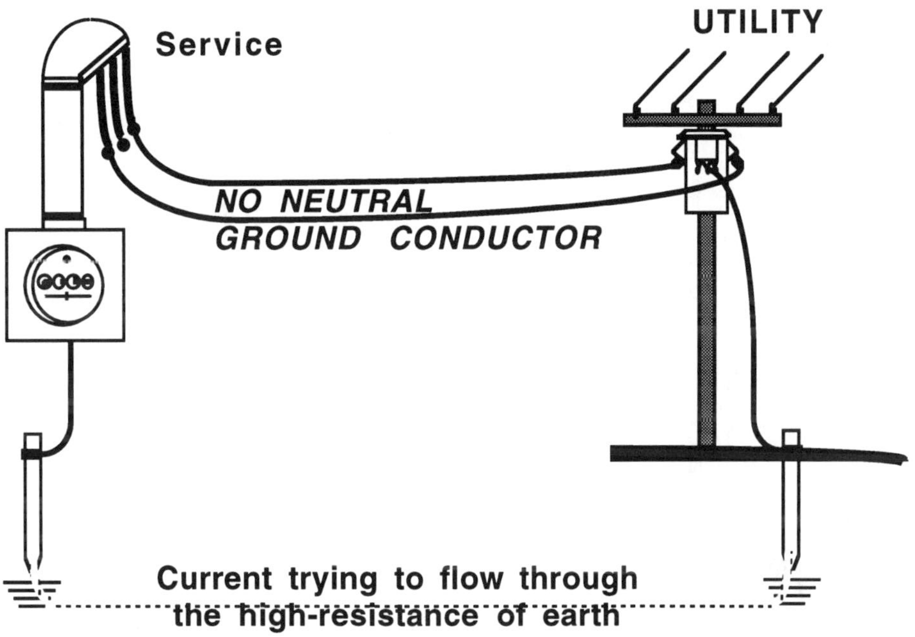

If the ground circuit conductor is not brought to the service equipment and a phase-to-ground fault occurs, the fault current path back to the source is on the equipment grounding conductor to the grounding electrode conductor, grounding electrode, earth, then through the grounding electrode conductor to the grounded terminal on the distribution transformer.

This high resistance path will limit the fault current to a value less than required to open the overcurrent protective device ahead of the fault. A condition now exists that can cause a fire and can create an electric shock hazard.

NO NEUTRAL LOADS

Whenever a service is derived from a grounded neutral system, the grounded neutral conductor must be brought into the service entrance equipment, even if there are no neutral loads connected to the panel. The neutral is required to provide a low impedance path for ground-fault current to return to the neutral to ensure operation of the fuse or circuit breaker.

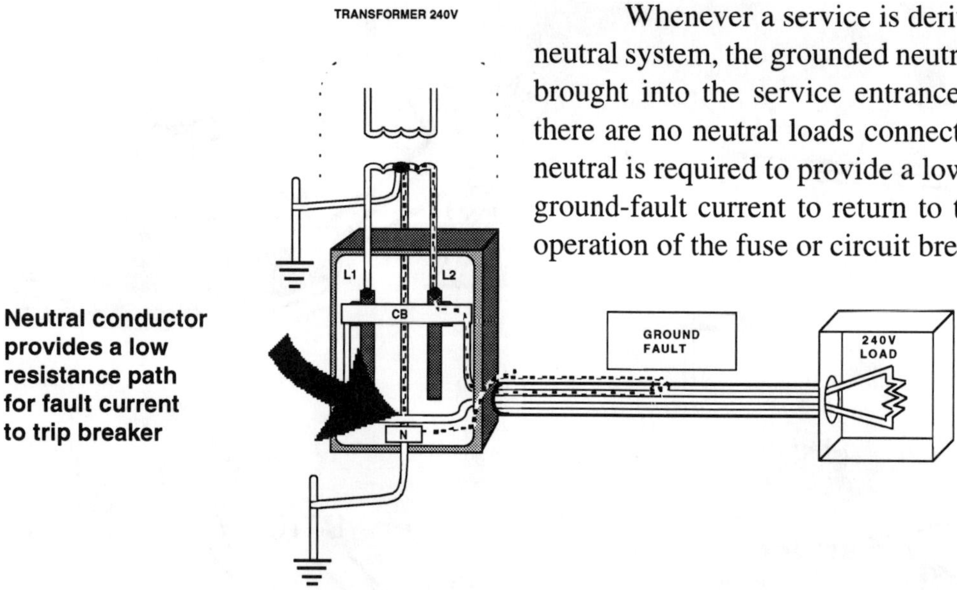

The neutral functions just like an equipment grounding conductor, to provide a path back to the transformer and to the overcurrent device to open the circuit that faulted. Without the neutral in this case, the fault current would have to flow through earth and would not open the overcurrent device.

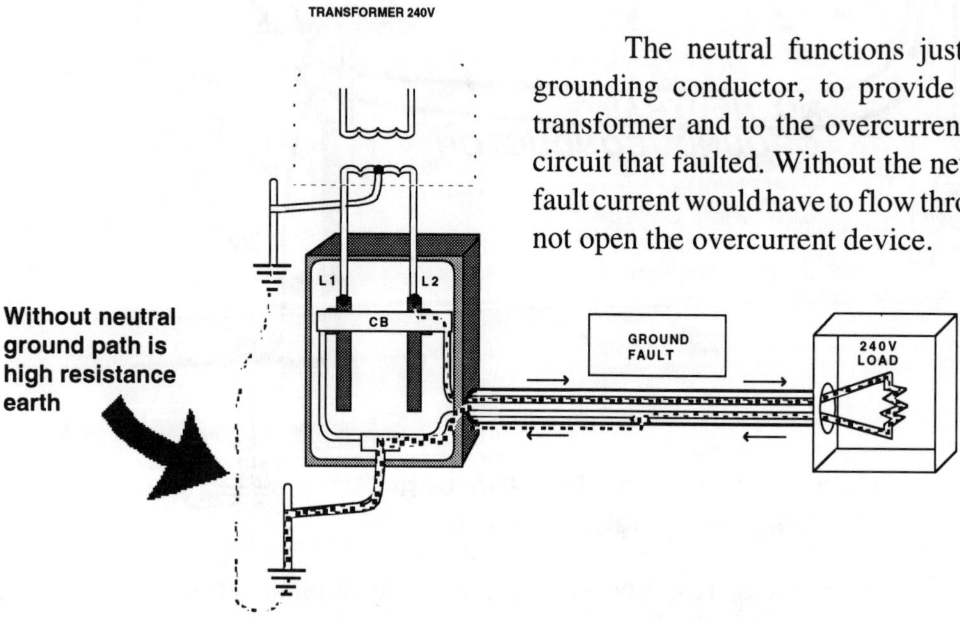

GROUNDED NEUTRAL CONNECTION

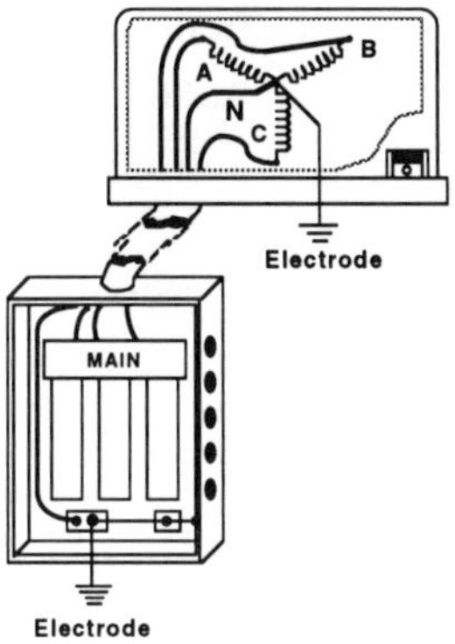

When an electrical system is grounded, a connection to a suitable grounding electrode is required at the service entrance. In addition to the grounding connection for the grounded (neutral) conductor at the point of service entrance to the premises, it is further required that another grounding connection be made to the same grounded conductor at the transformer that supplies the system.

The grounded (neutral) conductor connection to the grounding electrode must be anywhere from the *load end* of the service drop or lateral.

The grounding electrode conductor may be connected to either the neutral bus (or terminal lug) or to the equipment grounding block or bus in any system that has a conductor or busbar bonding the neutral bus or terminal to the equipment grounding block or bus. Where the neutral is bonded to the enclosure by a bonding *screw*, the grounding electrode conductor must *not* be connected to the grounding block or bus. It must be connected *only* to the *neutral conductor terminal block or bus.*

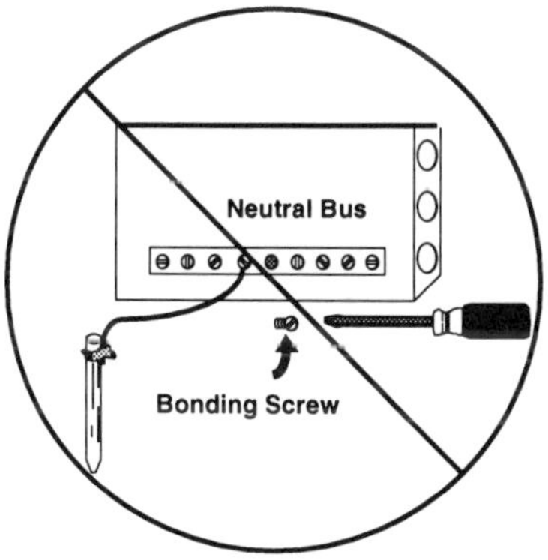

The reason is that the conductivity of a *screw-type* bond to the grounding system is not adequate to pass the high currents of say a lightning strike on the system.

The general rule is: Once the neutral is connected to a grounding electrode at the service equipment, the neutral must be *insulated* from all equipment enclosures and other grounded parts on the *load side* of the service.

The use of a neutral to ground, a subpanel, or other equipment on the *load side* of service equipment would be extremely hazardous if the neutral became loose or disconnected.

The bonding of subpanels or any other neutral connection to equipment enclosures is prohibited.

THE GROUNDING CONDUCTOR

The Code requires that metal boxes, cabinets, and equipment enclosures be grounded. The equipment grounding connection can be made by the *metal* raceway or by an equipment grounding conductor. An equipment grounding conductor *must* be installed in PVC raceway. When an equipment grounding conductor is used, it must be installed within the *same* raceway, cable or cord with the circuit conductors.

External grounding of equipment housings, frames or enclosures is a violation. The equipment grounding conductor must be run with the circuit conductors.

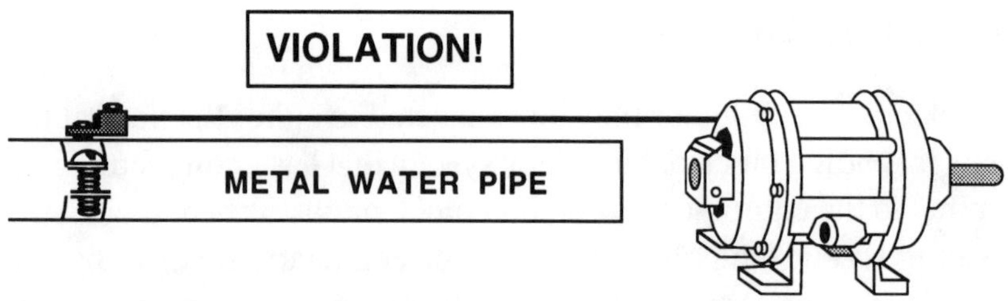

An effective equipment grounding system should, under conditions of *maximum* ground fault current flow, maintain a low potential difference on metallic enclosures and carry the maximum current without damage to the conductor or equipment. The grounding conductors must not burn off during a fault condition leaving the equipment "energized" creating a shock hazard.

Metal conduit is part of the electrical circuit, therefore the joints must be tight and clean to carry the large current under accidental fault conditions.

The Code places no restriction on the *size* or *length* of metal conduit. It must have continuity as well as the required conductivity to pass enough current to open the breaker or blow the fuse under a fault condition.

Joints in conduits and raceways require *special* consideration to avoid a shower of sparks that accompanies a fire hazard during a fault condition.

LOOSE LOCKNUTS

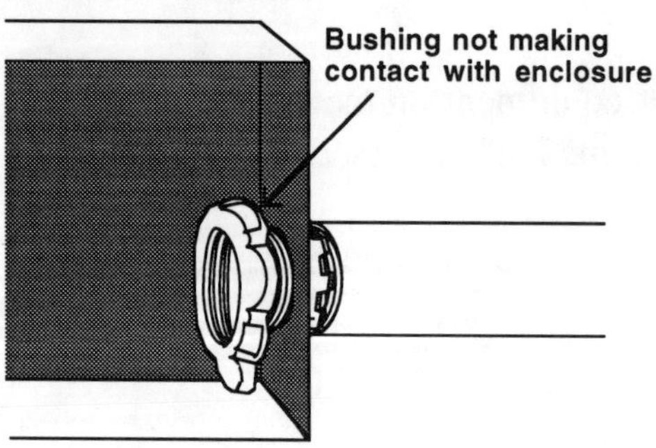

Some electricians know better but are careless in their work and this results in loosely assembled conduit systems which will definitely throw sparks if they are called upon to carry fault currents.

The calculation of impedance for metal conduit is laborious and inaccurate. When metal becomes part of an electric circuit, increases in both resistance and reactance will be large. Moreover, the resistance and reactance of the circuit will vary, depending upon the amount of current being carried.

Tests have shown that a metal conduit properly sized for the enclosed conductors will provide adequate capacity as an equipment grounding conductor. Although steel has a relatively high resistivity, it provides a large cross-sectional area. A 1/2" rigid steel conduit has a cross section of over 300 kcmil. The tubular shape and the fact that it is concentric with respect to the supply conductors it encloses contribute to reducing the overall impedance of the conduit. The equipment grounding conductor is *not* designed to carry current continuously, but to provide a low impedance path for fault current for a *short time* until the breaker opens or the fuse blows.

Even though the Code does *not* require an equipment grounding conductor installed within the metal conduit, it still is the best design for a number of reasons.

The safety of the grounding system will depend on how long we can expect the equipment grounding conductor to remain *permanent* and *continuous*.

Corrosion alone can open the continuous path on steel conduit and often a steel coupling is replaced with a PVC coupling. The replacement of equipment and not tightning the locknuts or fittings properly can interrupt the effective grounding path. As you can see, over a period of time, once which was a properly installed equipment grounding circuit can result in an ineffective path of current flow.

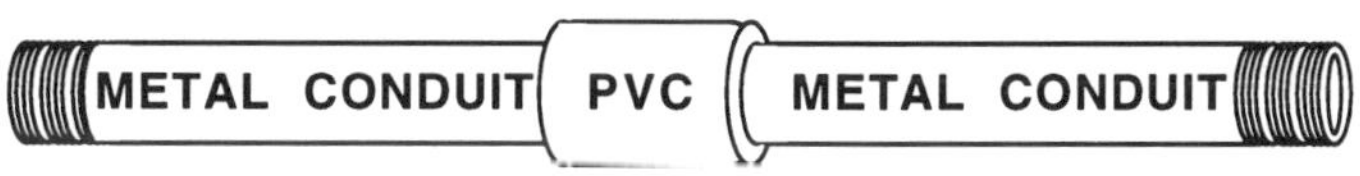

I have seen enough broken or corroded conduit to convince me that an equipment grounding conductor should be installed in **ALL** conduits. The Code recognizes metal conduit as an acceptable grounding path. The key is to properly install it, and maintain it.

Equipment grounding must be thoroughly reliable in order not to give a false sense of security.

Electrical shock and severe burns have been caused by conductor and insulation failures.

The sketch below shows a test circuit which has a single conductor installed in a steel conduit. The first ammeter reads 100 amps. The single conductor is *bonded* to the steel conduit at both ends providing a parallel path for current flow.

The steel conduit is actually carrying 97 amps and the single conductor only 3 amps. AC current flows to the outer edge of a conductor due to "skin effect". The conduit is the *outer edge* since they are connected in parallel. The conduit is actually more of a current-carrying conductor than the conductor itself.

This points out the importance of making the conduit couplings, fittings, locknuts, etc. *tight* to carry the current effectively in the grounding path when required.

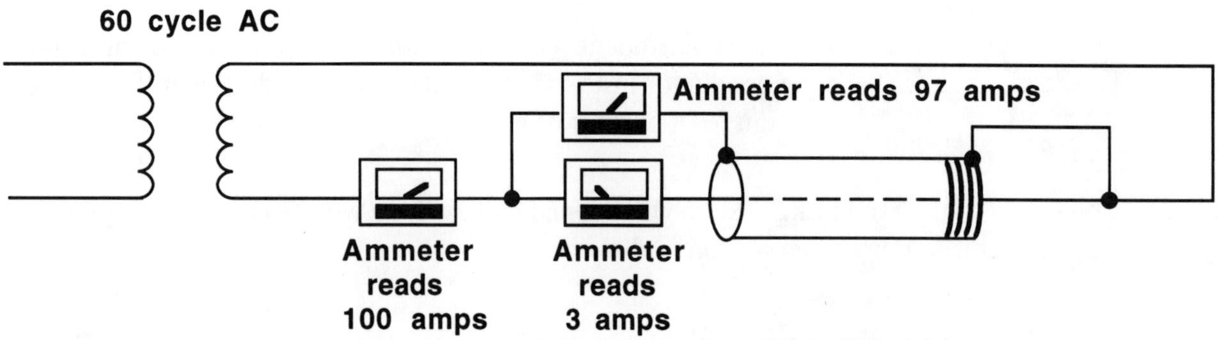

Thus, the steel conduit functions as a magnetic shield preventing the conductor within the conduit from carrying substantial current.

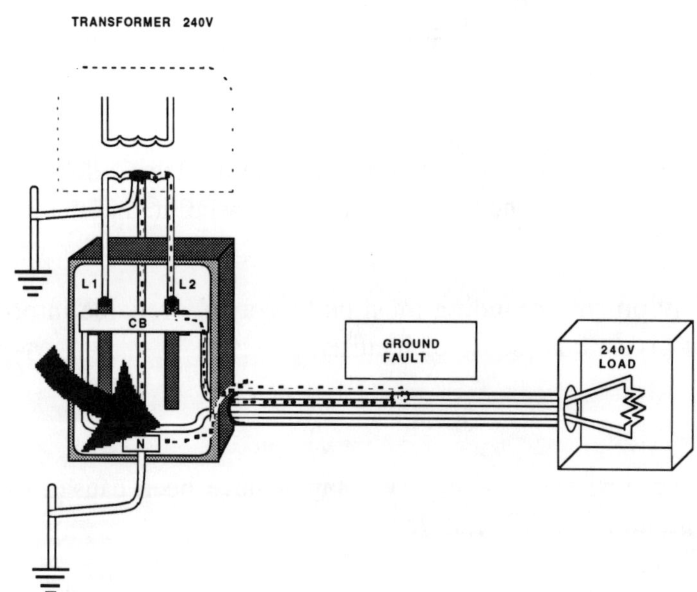

The equipment grounding conductor must always provide a low impedance path to open the breaker or fuse supplying the circuit conductors.

ELECTRICAL SHOCK

We've all experienced an electrical shock at one time or another. It's very important to understand how the human body reacts to an electrical shock.

A rule of thumb number for body impedance is 1000Ω. The skin impedance is a network of resistances, capacitances, voltages, frequency, duration of current, skin moisture, temperature, etc.

The route the current takes through the body affects the total impedance.

Blood has a very low resistance, almost equivalent to salt water.

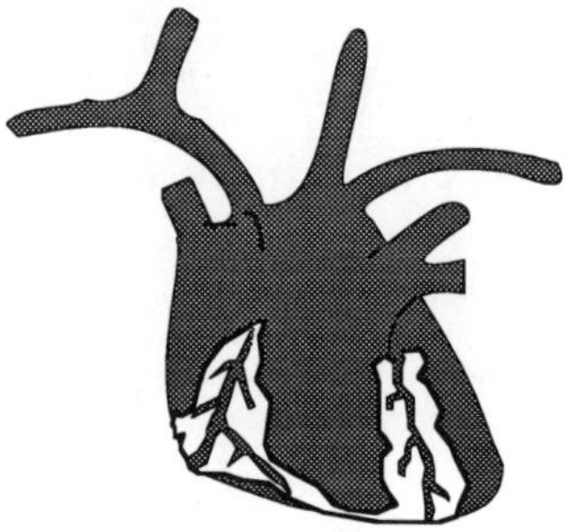

Your body generates a very small electrical current pulse that tells the heart when to beat.

The tiny natural body current is in the millionths of amps. A shock of one milliampere is a very strong shock. When an electric shock current passes through the heart it overwhelms the natural body signal. Even though the shock may be in milliamperes it is still thousands of times larger than the normal body current.

During an electrical shock the heart muscles get all mixed up as to what they are suppose to do, and just flutter, causing *ventricular fibrillation.* A heart can not over come this without help, a person will die within minutes from lack of oxygen in the blood supply.

Tests have determined that a human can not let go at the 30 milliampere level, you are frozen.

GFCI

A ground-fault circuit interrupter (GFCI), a people protector, can be either a circuit breaker or a receptacle.

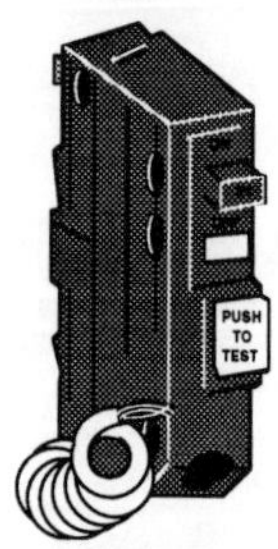

A GFCI has a built-in differential transformer which uses a magnetic field to measure the flow of current on the black wire and the flow of current on the white wire.

Under *normal* conditions, the current flow would be the same in both wires. With a ground fault leak to metal , the current in one wire **does not equal** the current flowing in the other wire. When this happens, the induction in the secondary winding of the transformer will cause the sensing unit to operate the GFCI.

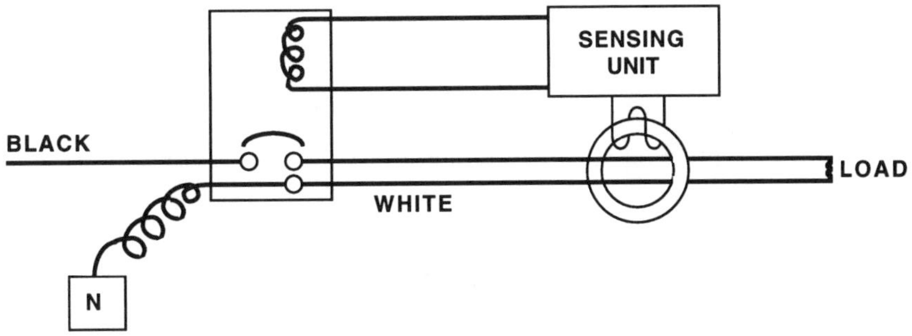

The GFCI will trip at **4 to 6 milliamperes**. The GFCI constantly monitors the flow of current in a circuit. If the current going out to the load differs by **.005 amperes** from the current coming back from the load the GFCI acts quickly and disconnects the circuit. Think about it for a moment, if the circuit current going out through a wire is 16 amps but the returning current is 15. 995 amps the circuit is shut off within 0.1 seconds. People purchase gold bathroom fixtures costing hundreds of dollars but question the $10 for a GFCI receptacle that can save their life.

The GFCI will **not** prevent a person from receiving a shock, but the shock will be minor.

As the name implies, ("Ground-fault Circuit Interrupter"), the GFCI does **not** protect a person that touches **both** wires at the same time. The device protects from ground-faults not short-circuits. Phase-to-phase short-circuit protection is the responsibility of the overcurrent device.

The first Code requirement for GFCI protection was in 1968 for swimming pool lights.

An equipment grounding conductor is not an effective people protector. The GFCI is the people protector from electrocution, grounding is no substitute for GFCIs.

If the body had an impedance of 1000Ω and a person touched a 120v wire, you would receive I = E/Z = 120v/1000Ω = .120 amps or 120 milliamps. It's the **time** a person is exposed to this current that can be fatal. The GFCI is to make the exposure time so short that ventricular fibrillation most likely will not occur.

Remember, no one can let go at 30 milliamps, so the GFCI is opening the circuit so you can let go.

The two ground fault shock conditions that can lead to electrocution are *series* and *parallel* contact.

A GFCI is extremely effective in preventing death in both conditions. An equipment grounding conductor is totally ineffective in series contact and the parallel contact effectiveness relies on several factors.

EQUIPMENT GROUNDING CIRCUIT

In series contact the person is the only path for current to flow. A person is using a metal hand held appliance with a 3-wire cord plugged into an adapter to a 2-wire receptacle. The black (hot) wire touches the metal frame of a hedge clipper. The person is the series contact to ground (earth).

SERIES CONTACT

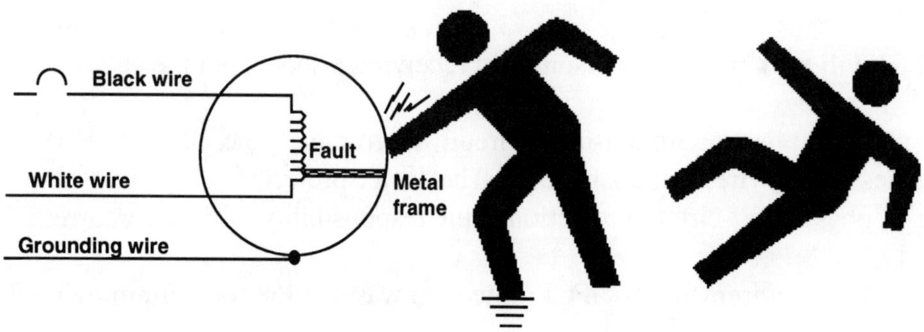

PARALLEL CONTACT

With the grounding conductor in the circuit it now becomes a parallel path with the person making contact to the fault to the metal frame. The amount of shock depends on the effectiveness of the grounding path and the speed in opening the overcurrent device. The current will divide in parallel with the majority of current following the path of least resistance. If metal conduit is used as an equipment grounding conductor and one of the fittings is not tight or missing, the path of least resistance will then be the person making contact.

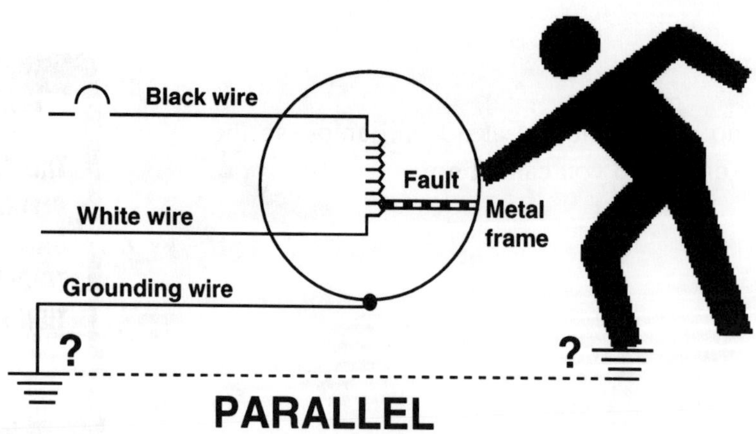

GFCI CIRCUIT

With a GFCI protector the person is protected with either a series or parallel contact as the sensing unit is detecting a differential of current between the black (hot) and the white (grounded) conductor. A leakage of 4 to 6 milliamps to the metal frame would open the GFCI overcurrent device within one tenth of a second.

SERIES CONTACT

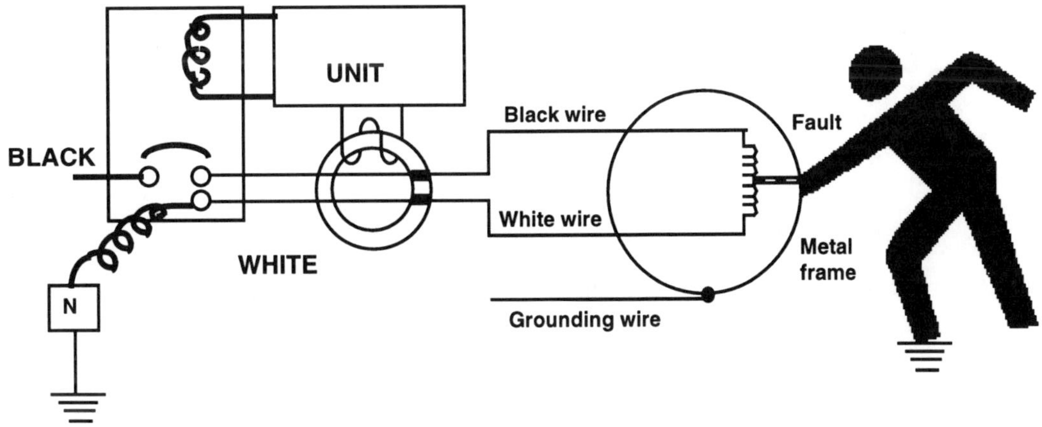

PARALLEL CONTACT

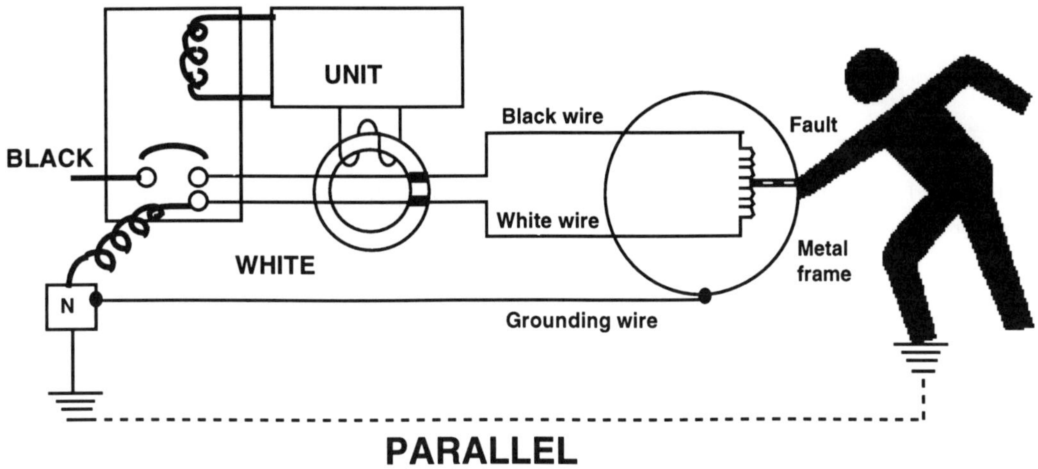

TOUCHING BOTH WIRES

A GFCI will **NOT** protect a person who touches both wires at the same time. There is no difference in current. The person is like a 25 watt light bulb to a GFCI.

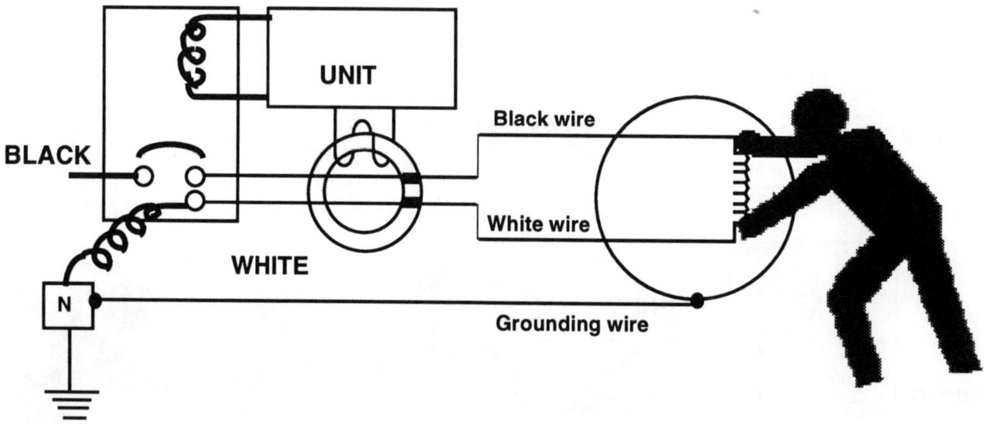

Remember we are talking about 15 mA causing muscular freeze and over 75 mA death.
15 mA = *.015 amperes* 75 mA = *.075 amperes*

A 25 watt @ 120 volt light bulb would have a current flow of 25w/120v = *.208 amperes or 208 mA*

These numbers far exceed the *75 mA fatal range.* The 208 mA current drawn from a 25w bulb could be fatal under certain conditions.

Again the level of current and shock depends on the conditions which involve the resistance.

A 25 watt @ 120 volt bulb has a resistance of 25w/.208a x .208a = *578 ohms*

A GFCI is continuously monitoring the grounded conductor on the load side. If the equipment grounding conductor would touch the grounded conductor the GFCI will trip as the circuit current is leaking through the grounding conductor.

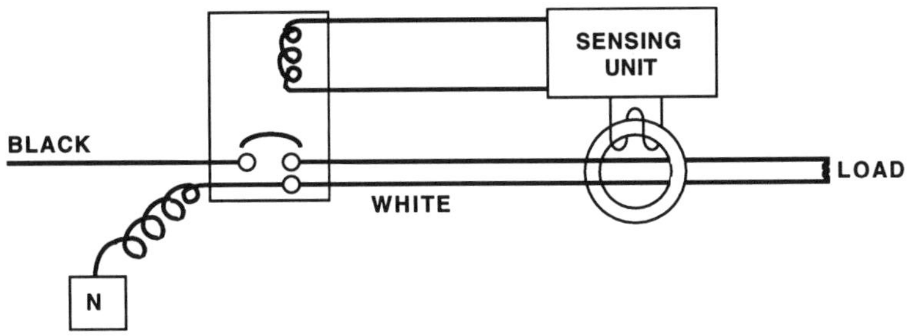

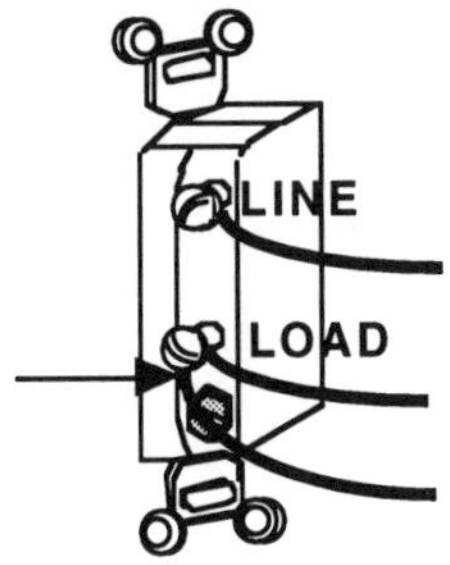

The concern was if a second ground occurs on the load side of the GFCI, ground fault current could go through a person, return to the circuit through the second ground through the GFCI sensing current transformer. The GFCI would not trip because there would be no difference in current. By adding a second current transformer it constantly monitors the grounded conductor looking for a second ground. Try touching the equipment grounding conductor to the grounded terminal of the GFCI receptacle and if wired correctly it will trip.

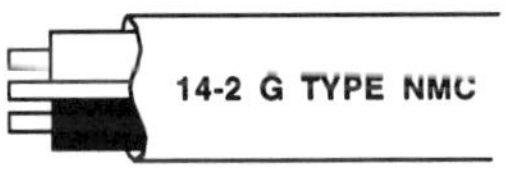

Driving uninsulated steel staples too tight against romex cable will cause a leakage of current and can trip the GFCI.

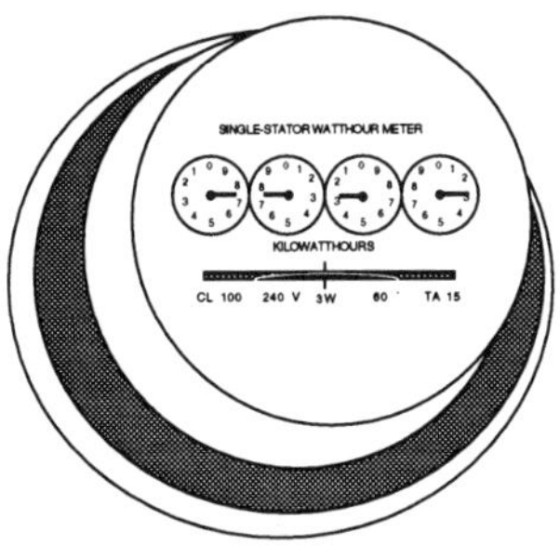

If the circuit wiring is not GFCI protected the current leakage wasting electricity by heating the steel staple could lead to a fire.

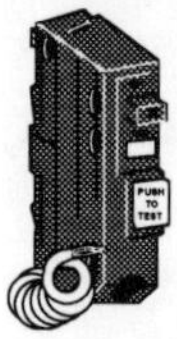

The advantage of a GFCI breaker is to protect the **entire** branch circuit, which includes the wiring in the walls.

The GFCI receptacle protects only what is plugged into it on the load side.

There are some appliances that have a *listed* standing leakage. Listed plug and cord connected appliances to be fastened in place or located in a dedicated space have a standing leakage of .75 milliamps. For portable plug and cord appliances the limit is .5 milliamps.

LEAKAGE .75mA

LEAKAGE .5mA

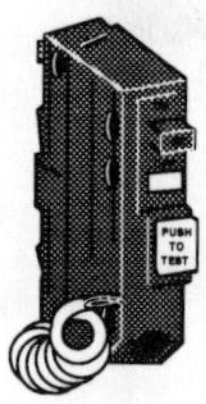

Remember the GFCI is designed to trip between 4 and 6 milliamps.

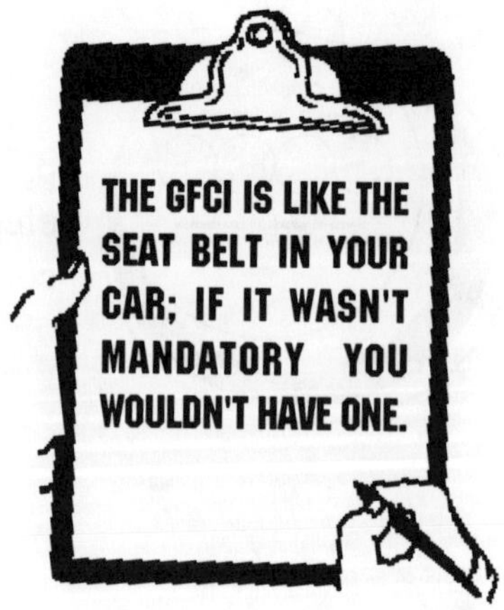

There is some listed equipment that allows a leakage current of 3.5mA. The higher leakage is allowed for FCC requirements by the use of EMI filters.

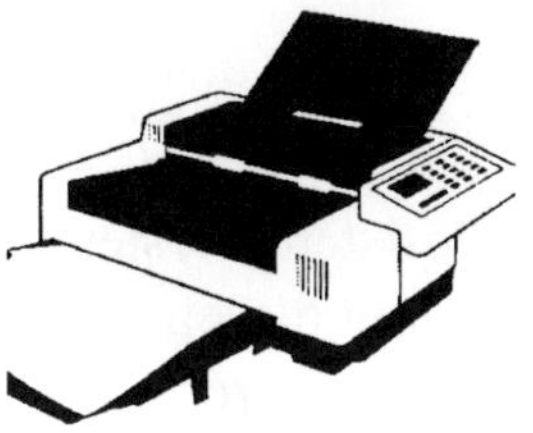

Some office machines are in this category.

When two or more of the pieces of equipment are plugged into the same GFCI protected branch circuit it would exceed the 6 mA and the GFCI would trip.

Proper grounding is confusing to many and this is easy for me to understand. Sometimes you'll find it just doesn't make any logic.

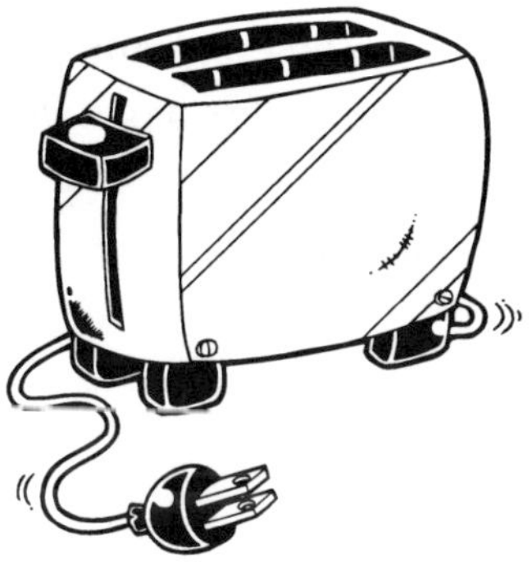

The Code requires an equipment grounding conductor to electrical equipment. The Code requires GFCI protection in the kitchen sink area, but yet we have metal appliances in the kitchen with a 2-wire cord and it's deemed okay?

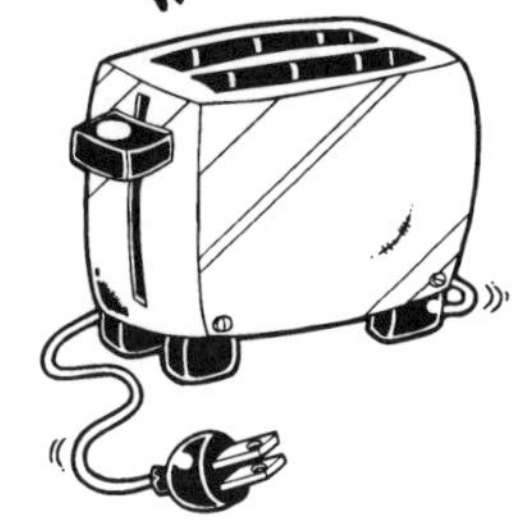

The logic for allowing a 2-wire cord with no equipment grounding conductor is if a piece of toast became lodged in the toaster and a person used a metal fork to retrieve the toast and the metal fork touched the bare heating element when it is energized and the metal edge of the toast slot, if the toaster were grounded with a 3-wire cord, the contact would cause a high level ground fault tripping the overcurrent device.

By using a 2-wire cord and not grounding the metal toaster, *as long as the person is not touching a grounded object* they will be okay to remove the toast that is stuck with the metal fork.

The argument is that grounding the toaster is not a good idea because if a person should touch the energized element with a metal fork and touched the grounded metal frame of the toaster at the same time, they could receive a fatal shock if it was not GFCI protected, which it is not required to be. That is the situation with any piece of equipment that is grounded.

Some facilities today, such as telecommunications buildings, require the ground resistance to be limited to a specific value. Thus, the only way to satisfy such a specification is to measure the ground resistance of the grounding electrode system.

To find the **ACTUAL** earth resistance of an electrode, an earth resistance tester must be used.

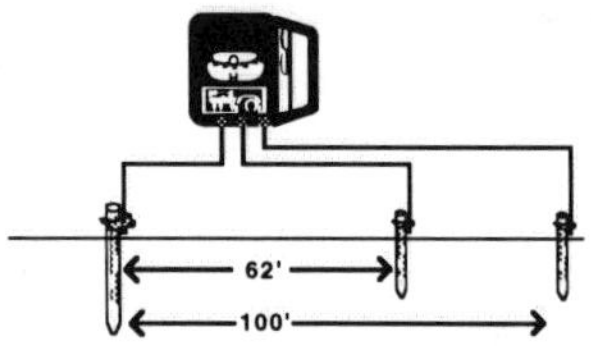

Tests were made to my residence using an earth resistance tester and the results were: The resistance to earth with one 8 foot driven ground rod was *345 ohms.* The resistance to earth with two ground rods driven 8 feet deep and connected in parallel was *150 ohms*, a reduction of 43%. Both rods connected to a metal underground water pipe was *3.8 ohms.*

You can see from the tests that a ground rod alone does not provide an earth resistance of 25 ohms or less required for a made electrode by the Code. By driving the second rod, it still did not lower the resistance to 25 ohms or less.

Because of its size, the earth is able to absorb an unlimited amount of current and for that reason, is called an unlimited conductor. Since earth is composed mostly of silica and alumina (both excellent insulators), its resistivity varies from 2 ohm-meters to 30,000 ohm-meters or greater.

If two 8 foot ground rods are driven they should be at least 16 feet apart to obtain the optimum effect.

A drop of 10° F below freezing increases earth resistivity 10 times.

Grounding electrodes are part of a ground reference arrangement that stabilizes the voltage to ground and dissipates surges. In performing this function, they are never intended to carry significant amounts of current for extended periods of time.

An insulation failure on a 480v circuit that energizes conductive materials that are connected to earth through an electrode, and with no other return path, current will flow through the electrode in accordance with Ohm's Law. If the electrode resistance is the Code maximum 25Ω, $I = E/R = 480v/25\Omega = 19$ amps of current flowing through the electrode. This most likely will NOT trip the circuit breaker. Now it becomes an extreme touch hazard.

The ideal condition is to have a grounding electrode with a resistance to ground so small that the voltage to ground would be held to a small value under any condition.

An earth resistance test would be an excellent addition to a local Code for safety.

To learn more about grounding I would recommend my book "The Grounding Workbook".

CHAPTER 4 EXAM

QUESTIONS ON GROUNDING - EXAM 1

• *Circle your answer.*

1. Planet earth is a relatively poor conductor.

(a) true (b) false

2. The ground rod actually plays no part in the tripping of a circuit breaker during a ground fault condition.

(a) true (b) false

3. Which of the following are made electrodes?

I. ground ring II. plate III. building steel

(a) I only (b) II only (c) II and III (d) I and III

4. A ground rod driven 8 feet into the earth is considered an excellent ground.

(a) true (b) false

5. The factors involved with resistance to current flow through an earth electrode system are ____.

I. the resistance of the ground rod itself and the connection to it
II. the contact resistance between the ground rod and the earth surrounding it
III. the earth resistance surrounding the ground rod

(a) I only (b) II only (c) III only (d) I, II and III

6. Increasing the diameter of the ground rod is a good way to decrease the earth resistance.

(a) true (b) false

7. Which of the following would improve the resistance to earth?

I. Use multiple ground rods
II. Treat the soil
III. Lengthen the ground rod

(a) I only (b) II and III only (c) I and III only (d) I, II and III

8. Chemical treatment is a permanent way to improve earth resistance.

(a) true (b) false

9. The Code uses ___ ohms as a maximum resistance limit for a single made electrode.

(a) 5 (b) 2.5 (c) 15 (d) 25

10. The best electrode for a low earth resistance would be a ____.

I. ground rod II. concrete-encased electrode III. plate electrode

(a) I only (b) II only (c) III only (d) I and III only

11. The Code requires rods of steel or iron to be at least ____ in diameter.

(a) 3/4" (b) 1/2" (c) 5/8" (d) 1"

12. The Code requires a ground rod to be driven a minimum of ____ feet into the soil.

(a) 7 (b) 8 (c) 7 1/2 (d) 6

13. No unguarded portion of the ground rod should be above earth for safety reasons.

(a) true (b) false

14. The connection of the grounding electrode conductor to the ground rod must be visible.

(a) true (b) false

15. Underground gas piping can be used for the electrode if at least 10 feet is buried.

(a) true (b) false

QUESTIONS ON GROUNDING - EXAM 2

• *Using the Code book circle answer and record Code section where found.*

1. The grounded service conductor shall not be smaller than the required ____.

(a) grounding electrode conductor **(b) largest phase conductor**
(c) ungrounded service conductor **(d) largest equipment conductor**
Code section __________.

2. Which of the following statements is (are) true?

I. Lightning rods shall be spaced at least 6 feet away from noncurrent-carrying metal parts of electrical equipment.
II. Lightning rods shall be bonded to the lightning rod conductors where separation distance is less than 6 feet.

(a) I only (b) II only (c) both I and II (d) neither I nor II
Code section __________.

3. Where extensive metal in or on buildings may become energized and is subject to personal contact ____ will provide additional safety.

(a) adequate bonding and grounding **(b) bonding**
(c) suitable ground detectors **(d) none of these**
Code section __________.

4. The equipment bonding jumper on the supply side of the service is sized by the rating of the ____.

(a) overcurrent protective device **(b) service entrance conductors**
(c) feeders **(d) load to be served**
Code section __________.

5. The #4 solid copper conductor connecting the grounding electrode to the service has been accidentally cut. Which of the following would be acceptable connectors for splicing this conductor?

I. a split bolt connector II. a compression sleeve type connector III. a wire nut

(a) I only (b) I or II only (c) I, II or III (d) none of these are acceptable
Code section __________.

6. ____ and larger grounding electrode conductors shall be protected where exposed to severe physical damage.

(a) #8 (b) #4 (c) #2 (d) #6
Code section __________.

7. Where conductors are adjusted to compensate for voltage drop, equipment grounding conductors, where required, shall be adjusted proportionally according to ____.

(a) diameter (b) cross section area (c) circular mil area (d) circumference
Code section __________.

8. A ground ring electrode must be buried a minimum of ____.

(a) 12" (b) 18" (c) 24" (d) 30"
Code section __________.

9. A concrete-encased electrode must be encased with at least ____ inches of concrete.

(a) 2 (b) 4 (c) 6 (d) 8
Code section __________.

10. A concrete-encased electrode located near the bottom of the foundation shall have at least one rebar ____ feet long and a minimum of ____ " diameter.

(a) 10 - 1/2 (b) 15 - 5/8 (c) 20 - 1/2 (d) 12 - 3/4
Code section __________.

11. Where connected to a concrete-encased electrode, that portion of the grounding electrode conductor, which is the sole connection to the grounding electrode, shall not be required to be larger than ____ copper.

(a) #6 (b) #4 (c) #2 (d) #8
Code section __________.

12. A green colored hexagonal screw on a duplex receptacle is used to connect the ____ conductor.

(a) grounded (b) ungrounded (c) grounding (d) isolated
Code section __________.

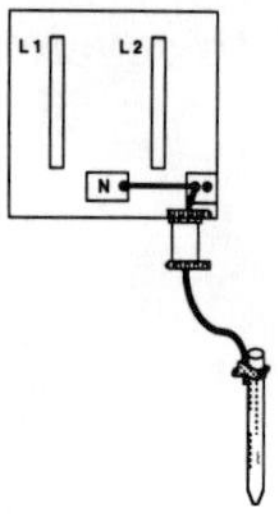

13. The sketch shows the grounding electrode conductor from the grounding bus to the ground rod being protected by a rigid metal conduit bonded at the service enclosure by a grounding bushing.

☐ OKAY ☐ VIOLATION

Code section ________.

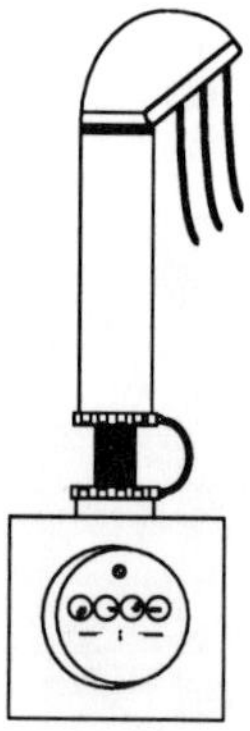

14. Service entrance cable is installed inside a rigid steel mast. The Code requires a bonding jumper from the mast to the meter enclosure to maintain an effective grounding path. The service is 150 amp 240/120v single-phase supplied with size #1/0 THW copper conductors. The grounding electrode conductor to the electrode is a #6 bare copper. The bonding jumper for the service mast is also a #6 copper.

☐ OKAY ☐ VIOLATION

Code section ________.

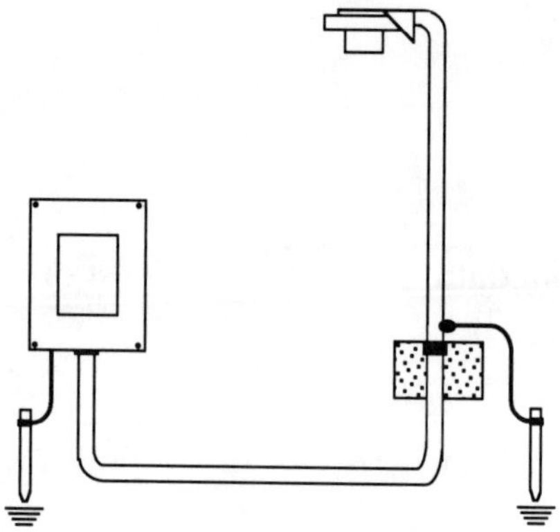

15. A 20 amp branch circuit feeds a 250 watt mercury-vapor light on a metal pole. The circuit is fed underground using PVC conduit to a junction box at the base of the metal pole. Since the light and the metal pole are grounded at the base of the pole with a ground rod driven 8 feet deep, no equipment grounding conductor is run with the circuit conductors.

☐ OKAY ☐ VIOLATION

Code section ________.

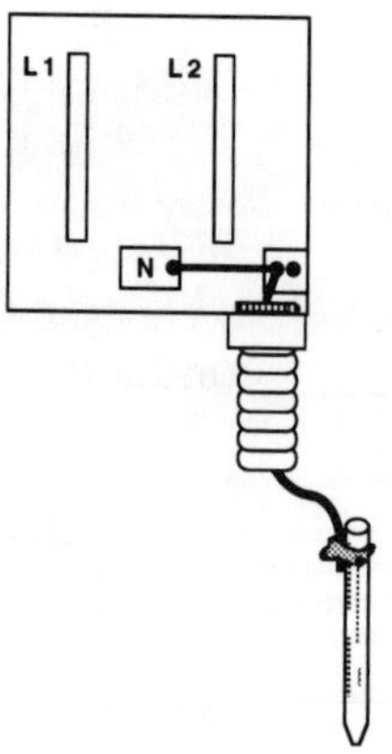

16. A short piece of metal flex conduit is used for mechanical protection for the grounding electrode conductor.

☐ OKAY ☐ VIOLATION

Code section ________.

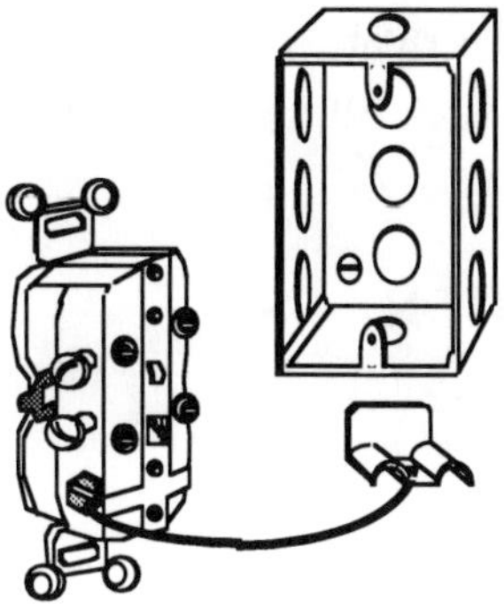

17. The equipment bonding jumper to the grounding pin on the receptacle is connected to the screw for the cable clamp.

☐ OKAY ☐ VIOLATION

Code section ________.

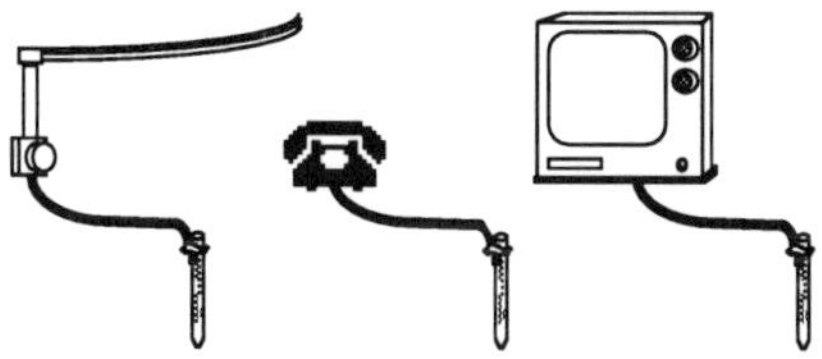

18. The service, telephone, and cable TV system all have separate ground rod electrodes that are not bonded together.

☐ OKAY ☐ VIOLATION

Code section ________.

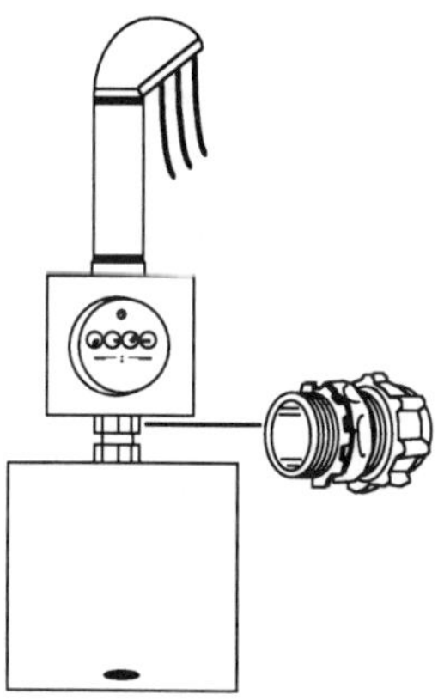

19. The service entrance contains threadless connectors for the EMT raceway between the meter enclosure and the service enclosure.

☐ OKAY ☐ VIOLATION

Code section ________.

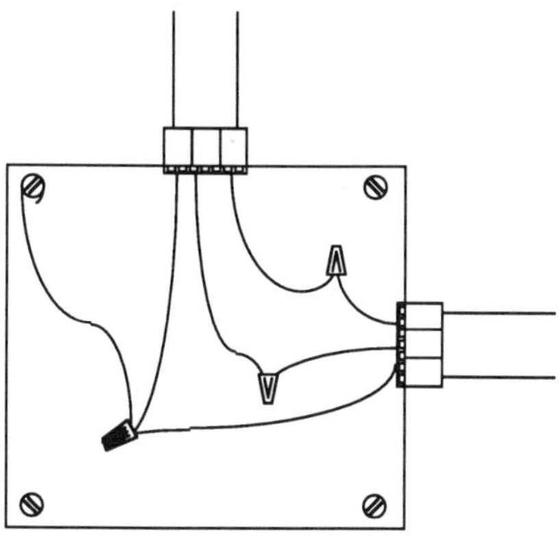

20. The equipment grounding conductor is connected to a mounting screw for the metal enclosure.

☐ OKAY ☐ VIOLATION

Code section ________.

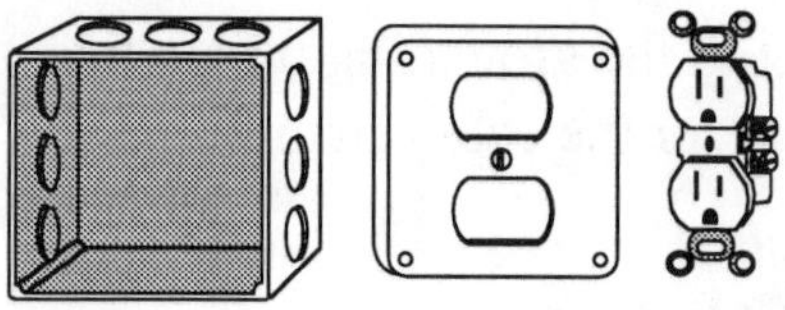

21. The duplex receptacle is attached to the raised box cover by a single #6-32 screw. There is no bonding jumper from the receptacle to the metal box.

☐ OKAY ☐ VIOLATION

Code section ________.

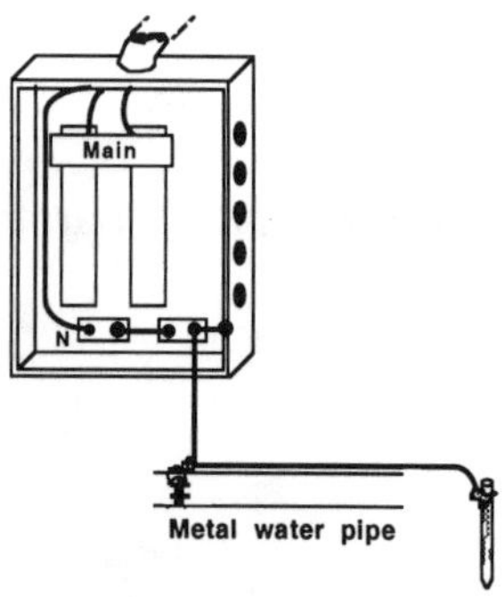

22. A service to a residence is 240/120v, single-phase, 3-wire, 200 amps with #2/0 THW copper conductors. What size copper grounding electrode conductor is required from the service equipment to the metal water pipe? # __________.

What size copper conductor is required from the metal water pipe to the ground rod? # __________.

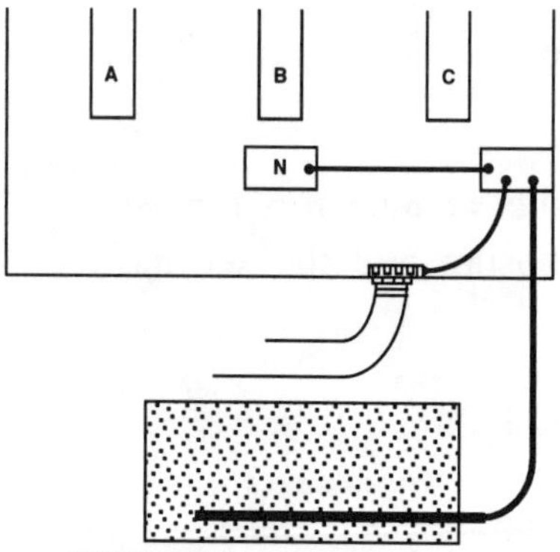

CONCRETE-ENCASED ELECTRODE

23. The service conductors are #500 kcmil copper per phase. The equipment grounding conductor is the sole connection to the concrete-encased electrode. What size copper grounding electrode conductor is required? # __________.

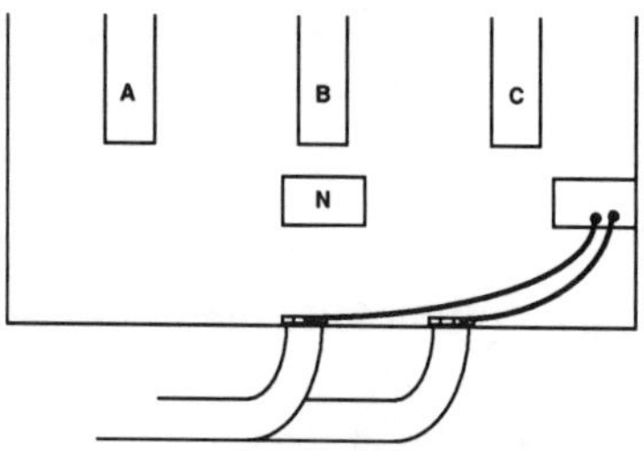

24. Two - #400 kcmil THW copper conductors are paralleled per phase. The overcurrent device protecting the circuit is 500 amperes. What size copper equipment grounding conductor is required for each PVC conduit? # __________.

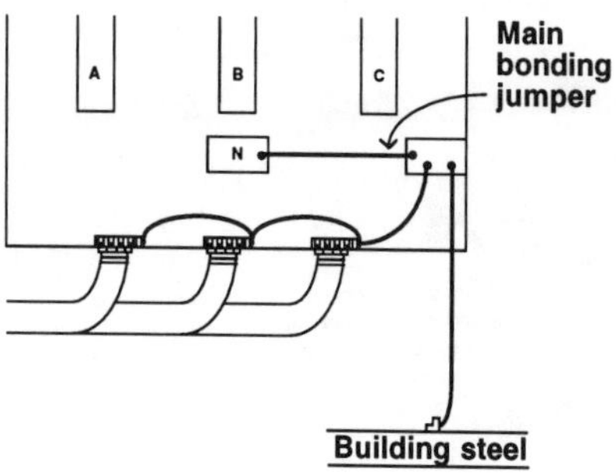

25. The service entrance contains 3 - #500 kcmil copper conductors paralleled per phase. What size copper main bonding jumper is required? # _________. What size copper equipment bonding jumper is required between conduits? # __________.

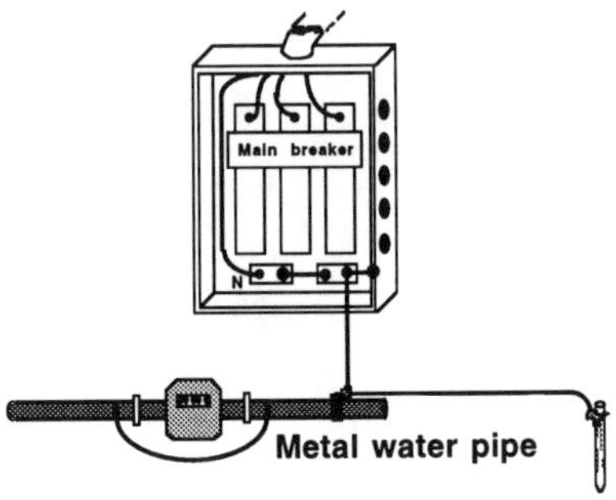

26. A three-phase service has #500 kcmil THW copper service entrance conductors. What size copper bonding jumper is required to bond around the water meter? # _________.

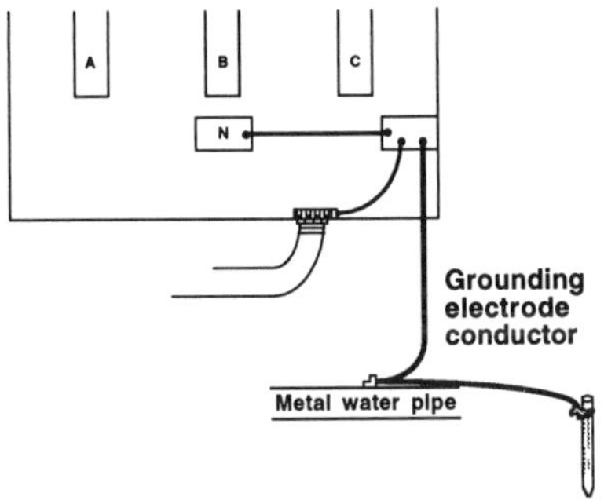

27. The service consists of #400 kcmil THW aluminum per phase. The main breaker is 250 amperes. What size **copper** main bonding jumper is required? # __________.

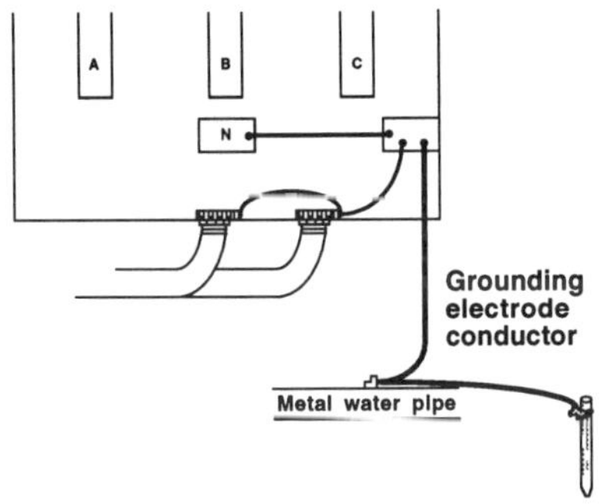

28. A three-phase service has 2 - #600 THW copper conductors paralleled per phase. What size copper grounding electrode conductor is required from the service equipment to the water pipe? # __________.

What size copper conductor is required from the water pipe to the ground rod? # __________.

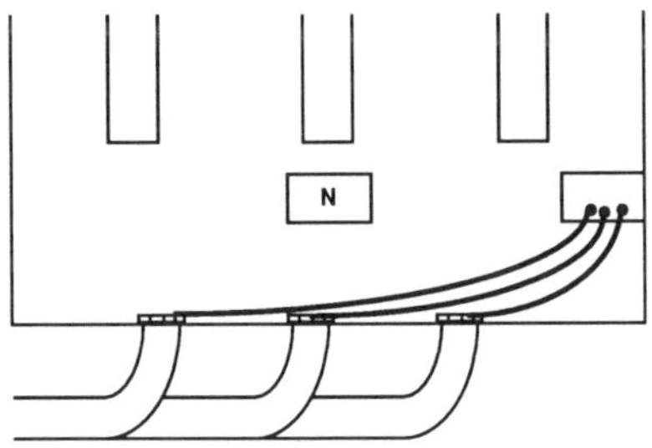

29. This sketch shows a feeder with 3 - #500 kcmil THW copper paralleled per phase. The feeder conductors are protected by a 1000 amp circuit breaker. What size copper equipment grounding conductor is to be installed in each PVC conduit? # ________.

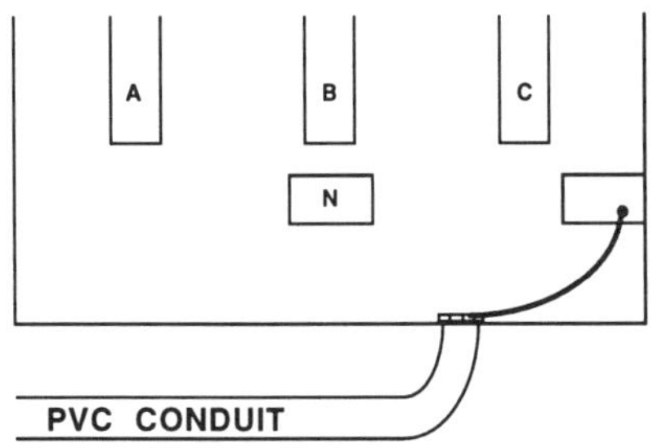

30. A feeder to a 400 amp distribution panel had the #600 kcmil conductors increased to #750 kcmil THW copper to allow for voltage drop. The larger feeder conductors are protected by a 400 amp circuit breaker. What size copper equipment grounding conductor is required for the PVC conduit? # __________.

CHAPTER 5

PLAN READING

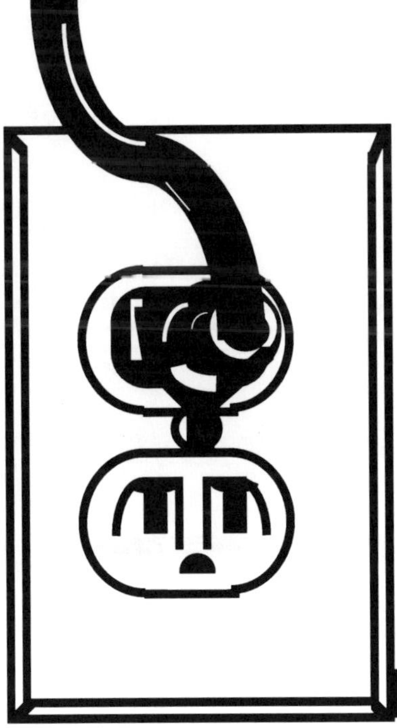

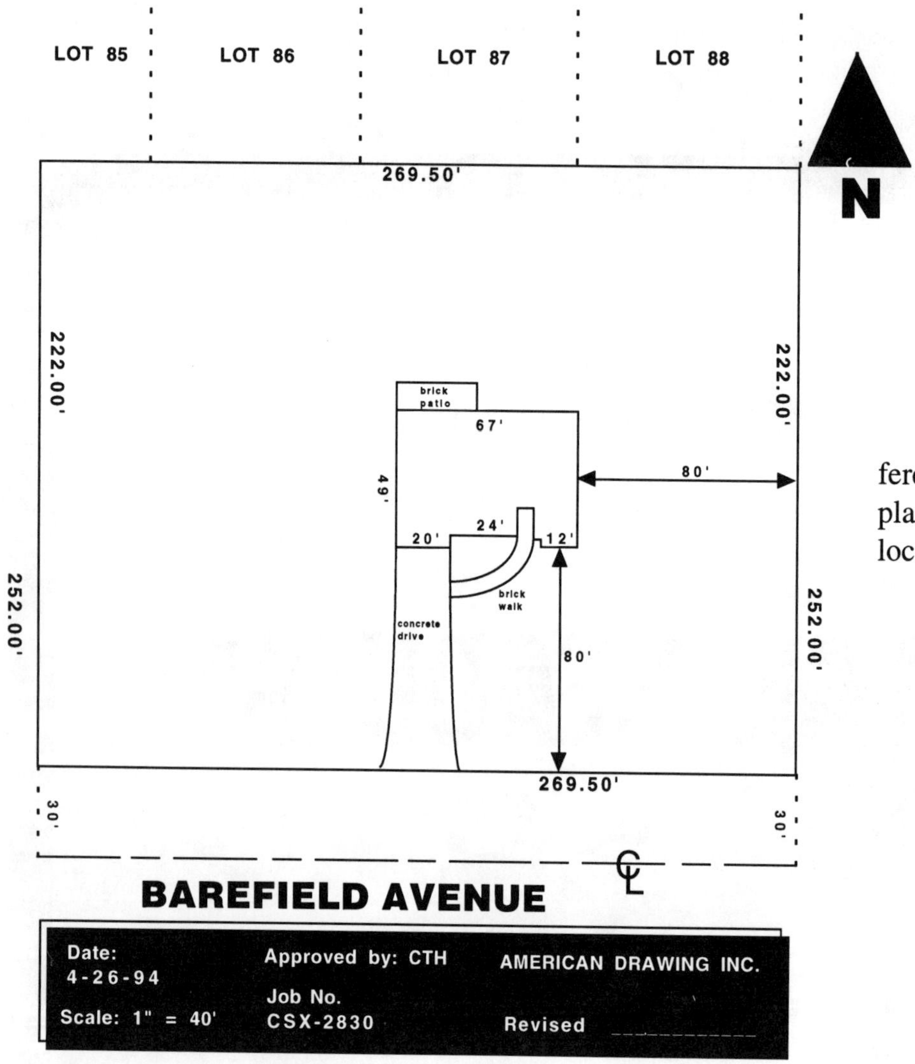

PLAN READING

There are many different type plans. The site plan shows how the house is located on the property.

The pictorial drawing shows the perspective of a building.

PERSPECTIVE VIEW OF HOUSE

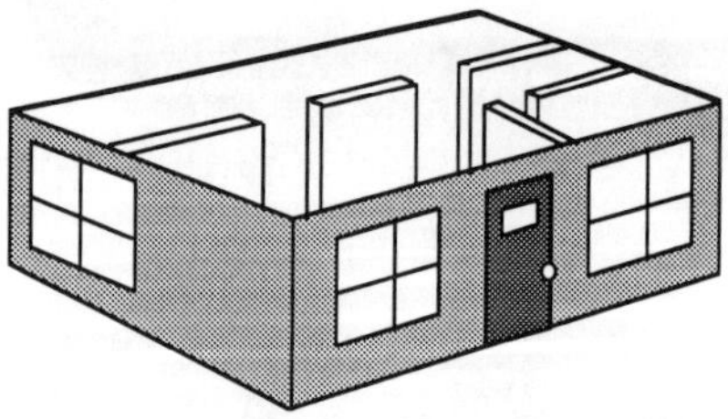

TOP HALF OF HOUSE REMOVED

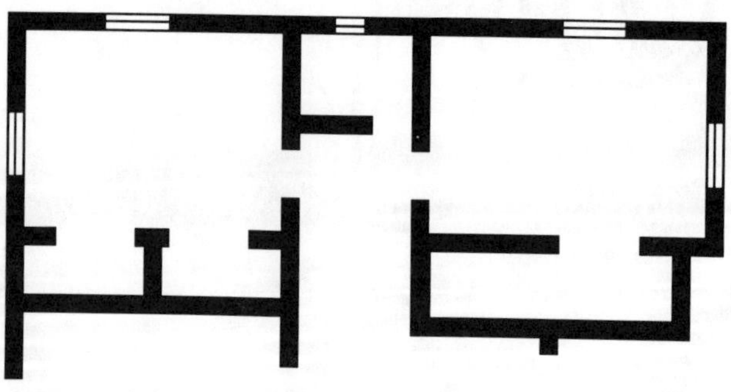

THE FLOOR PLAN IS VIEWING THE HOUSE FROM ABOVE

The wiring diagrams that you will use for constructions jobs are called blueprints. These diagrams use symbols to indicate components and to designate the specifications for various connections.

The plans examiner checks the plans and specifications for Code compliance before the permits are issued. An approved set of plans are kept on file at the building department.

Blueprints are drawn to scale, the most commonly used are 1/4" = 1 foot or 1/8" = 1 foot.

The blueprint is also called a floor plan in construction as it is a layout of a room, house, or building. A floor plan is a scaled, top view showing inner and outer walls with windows and doors also indicated. No furniture is shown in a floor plan, but major appliances such as ranges, dishwashers, and clothes dryers are shown on some drawings. Most floor plans include the symbols for sinks, toilets and bath tubs.The engineer or architect draws in special symbols representing light fixtures, receptacles, switches, and other components.

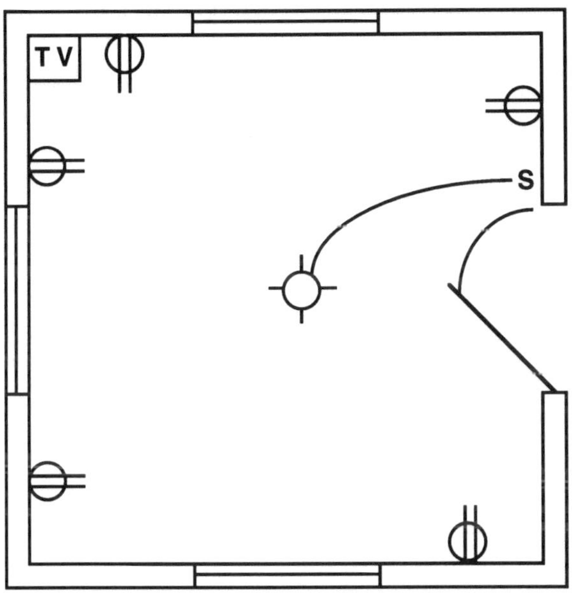

The floor plan does not show how the devices are connected. The electrician is suppose to know how to connect the devices. The electrician's responsibility of reading a blue print is to translate the floor plan into actual electrical wiring.

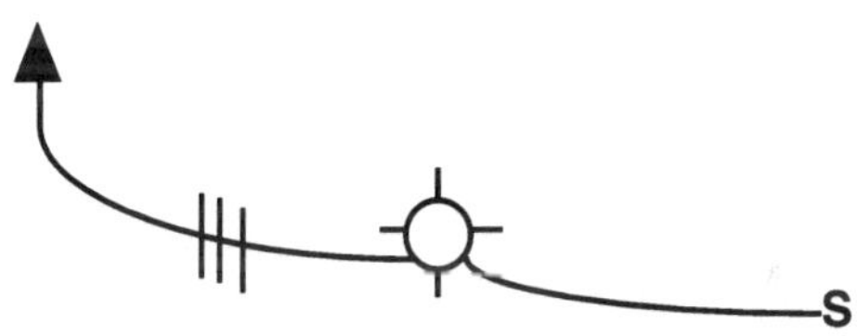

One line diagrams use special symbols for components. The components are shown interconnected by single lines. A single line can represent several wires.

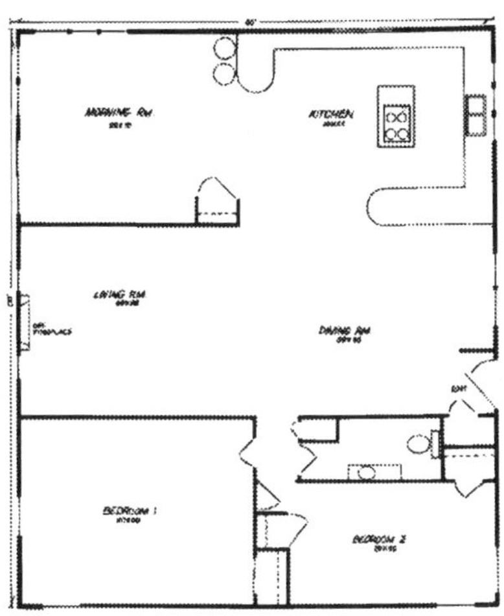

On a one-line diagram, the components, their exact location, and the wiring is shown. These drawings do not show specifications, cable or conduit type, or other supplementary information. This is usually provided on the blueprint in a separate specification box or table.

The two key parts of a blueprint are the symbols used to represent the components and the lines that represent the interconnecting wires.

Abbreviations are used on blueprints to indicate various components or aspects of the job.

ABBREVIATIONS and SYMBOLS

Abbreviations are used in electrical work to indicate various components or other aspects of the job. They can be used in the drawing, the bill of material, the specifications, or other written documents.

PNL = panel **SERV** = service **MTR** = motor **XFMR** = transformer **CKT** = circuit

Symbols are divided into five basic categories:

1. **GENERAL OUTLETS**
2. **CONVENIENCE RECEPTACLES**
3. **SWITCHES**
4. **PANELS, CIRCUITS, and MISCELLANEOUS COMPONENTS**
5. **AUXILIARY SYSTEMS**

Sometimes designers of blueprints modify the standard symbols to suit their own needs. For this reason, most drawings will have a symbol list or legend.

General outlet refers to a type of outlet or junction box. It can be for a switch, light fixture, etc. The symbol is a circle around a letter designating the specific type of outlet. The symbol with a short line extending out of the circle on the left side, indicates an outlet that is wall mounted. The symbol without the short line is used for outlets mounted in the wall.

wall **outlet**

Rectangular symbols are used to represent fluorescent light fixtures. You must also understand what the different letter designations mean. The letters indicate the type of outlet involved. A clock outlet has a "C", a pull switch "S", recessed fixture "R".

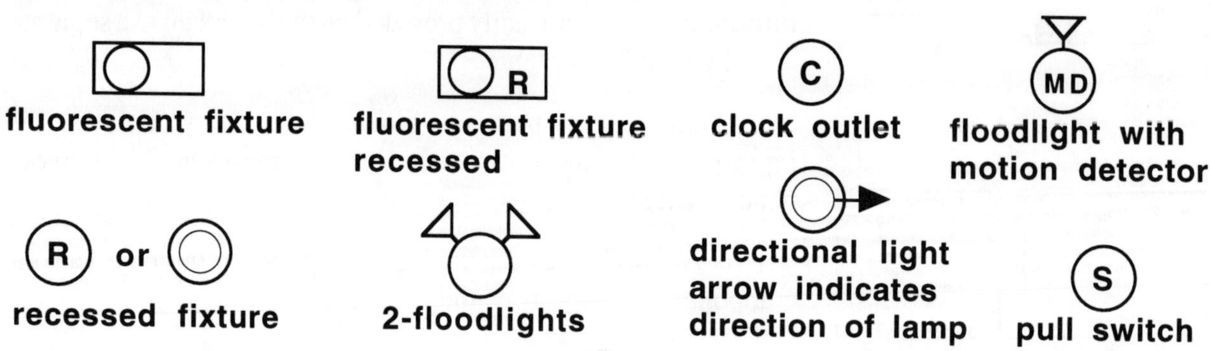

The symbol used for the receptacles is also a circle.

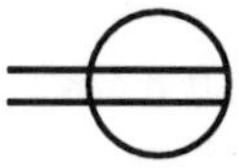

duplex receptacle

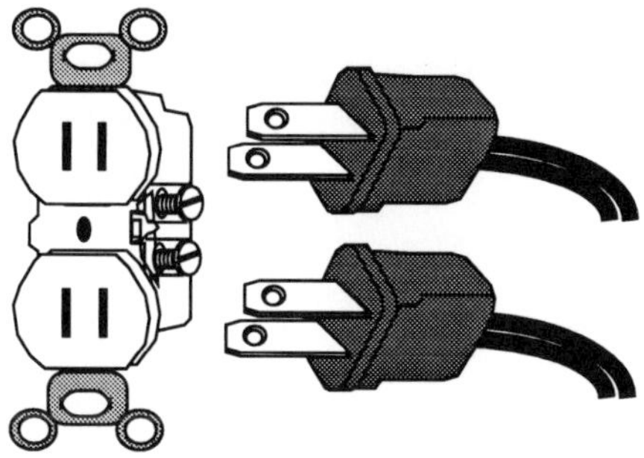

The most popular receptacle used is the duplex receptacle which accepts two cord connections. The symbol is a circle with lines. The lines indicate the type of receptacle: One line = single, two lines = duplex, three lines = triplex.

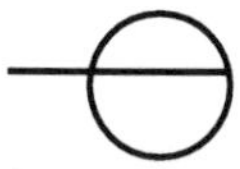

single receptacle

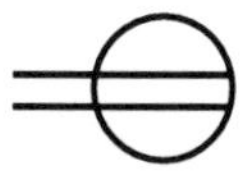

duplex receptacle

triplex receptacle

Sometimes a duplex receptacle is split so that one outlet is hot while the other outlet is switched. This symbol has a portion of it shaded. A duplex receptacle in the floor is shown in a box.

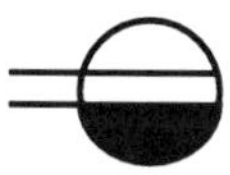

split-wired duplex

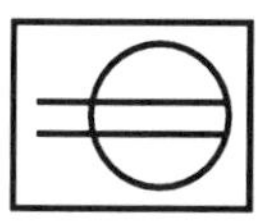

duplex receptacle in floor

end of run (last receptacle)

Special-purpose receptacles have a circle with a triangle in the center. These symbols indicate 240 volt appliance outlets such as ranges, clothes dryers etc. If the triangular portion of the symbol is open, a standard receptacle is indicated. When the triangle is darkened, a special-purpose connection is required. This means that the appliance is normally attached with a special connector rather than with a separate plug and convenience outlet.

single special-purpose receptacle

RH

special-purpose range hood

GD

special-purpose garbage disposal

R

range receptacle

For switch outlets on a blueprint a large **S** is used. Subscripts are used with the "S" to designate the type of switch required. The everyday common single-pole switch is shown with the letter "S" and no other marking. A three-way switch would have the letter "S" with a subscript "3". A four-way switch would have the letter "S" with a subscript "4".

S_3
three-way switch

S_4
four-way switch

The Code defines a lighting and appliance panelboard of one which has more than 10% of its overcurrent devices rated 30 amps or less, for which neutral connections are provided. This would be the case in house wiring, the panelboard would be a lighting panel. In the industry a panelboard with three-phase loads and motors without neutral connections would be an example of a power panel.

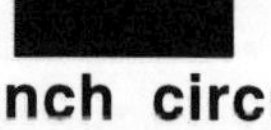

branch circuit lighting panel

distribution power panel

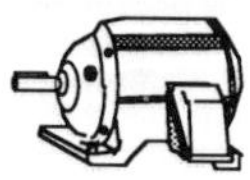

Other symbols often used are for motors and instruments. These are circular symbols with letter designations in the center.

motor motor

Meters are also designated with a letter symbol to indicate what type of meter.

watthour meter ammeter voltmeter power factor meter

Lines are used to indicate actual wiring interconnections. A solid straight line represents wiring that is concealed in a wall or ceiling. A line with long, wide dashes or alternating short-and-long dashes is used to represent wiring that is concealed in a floor. A dashed line made up of short dashes is used to represent exposed wiring.

CONCEALED IN WALL OR CEILING CONCEALED IN FLOOR EXPOSED

The dashed line can represent two things: Exposed wires or wires that run between a component and a switch are often used on a blueprint. They are usually shown as a curved, dashed line so they are not confused with architectural details which are represented by straight lines.

A solid line with an arrow indicates a branch circuit home run to the panel. A line with two arrows would represent two branch circuits.

The slashes in a line represent the number of wires in the cable or conduit. No slashes = one wire, one slash = one wire, two slashes = two wires, three slashes = three wires, etc.

3-wire cable

•Note: This does NOT include the equipment grounding conductor which the Code requires. This conductor is omitted to simplify the drawing.

ELECTRICAL SYMBOLS

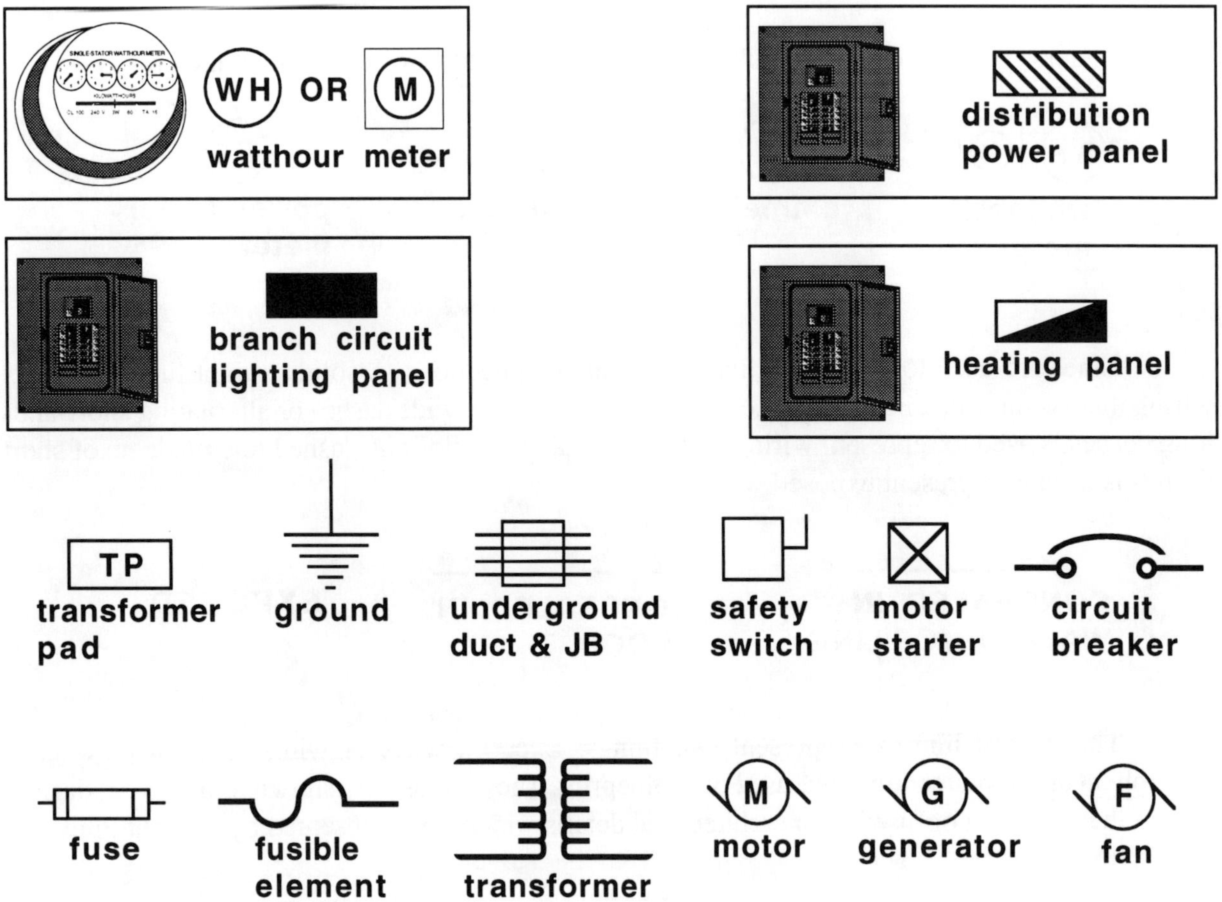

electric meters or instruments
replace · with identification letters:
A = ammeter V = voltmeter
MA = milliammeter VA = volt-ammeter
W = wattmeter WH = watthour meter
OM = ohmeter PF = power factor

WIRING

WIRING			
NOT CONNECTED	CONNECTED	CONTROL	POWER
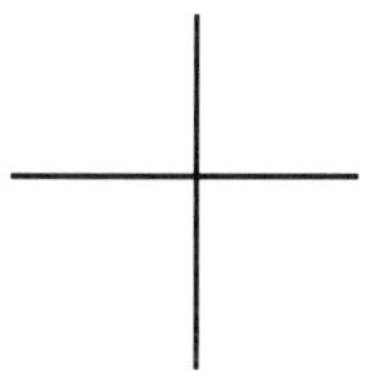	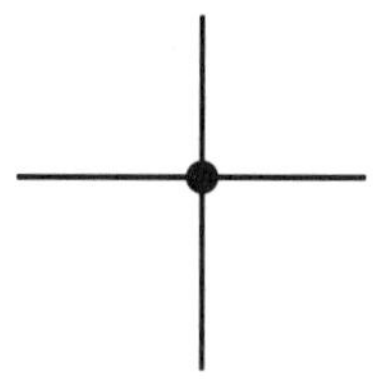		
HOME RUN	UNDERGROUND	CONCEALED	NUMBER OF CIRCUIT CONDUCTORS (4)

WIRING			
CONCEALED IN CEILING OR WALL	CONCEALED IN FLOOR	EXPOSED	WIRING TURNED UP

WIRING TURNED DOWN	CONDUIT ONLY	CONDUIT TURNED UP	CONDUIT TURNED DOWN

ELECTRICAL SYMBOLS

GENERAL OUTLETS

○	outlet
–○	wall
⊕	ceiling
Ⓙ	junction box
Ⓑ	blanked outlet
Ⓡ	recessed fixture
Ⓕ	fan outlet
Ⓢ	pull switch
Ⓒ	clock outlet
Ⓓ	drop cord

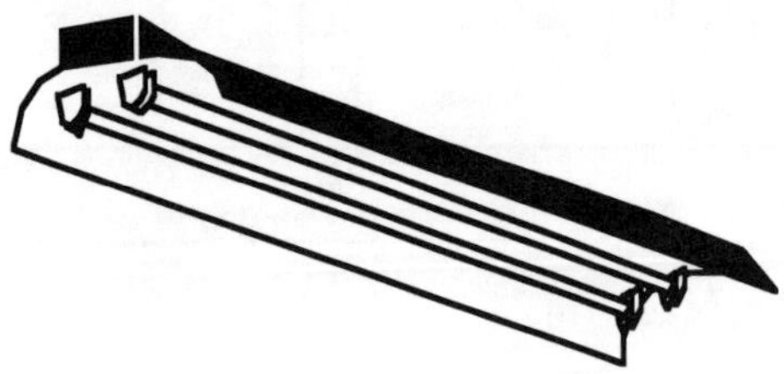

LIGHTING OUTLETS

▭ with ○	fluorescent fixture
▭▭▭ with ○	continuous-row fluorescent
├─┼─┼─┤	bare lamp fluorescent
Ⓥ	vapor discharge lamp
Ⓛ	lamp holder
Ⓛ PS	lamp holder with pull switch
○	incandescent
Ⓧ or ⊗	exit light outlet

ELECTRICAL SYMBOLS

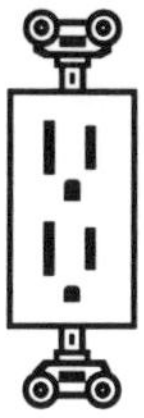

CONVENIENCE RECEPTACLES

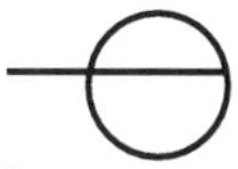
single receptacle

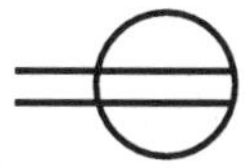
duplex receptacle

triplex receptacle

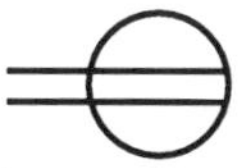
3
triplex receptacle

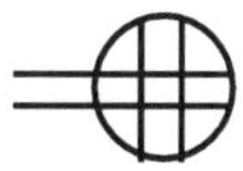
quadruplex receptacle

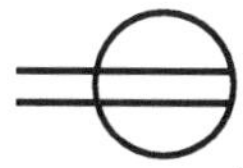
UNG
ungrounded duplex receptacle

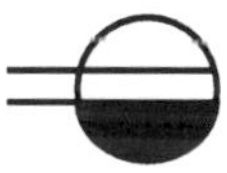
split-wired duplex

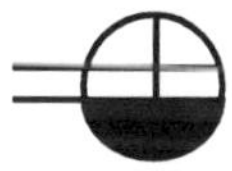
split-wired triplex

single special-purpose receptacle

duplex special-purpose receptacle

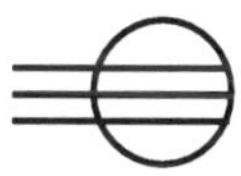
R
range receptacle

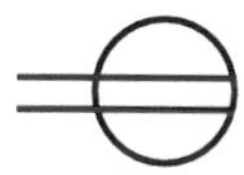
WP
weatherproof receptacle

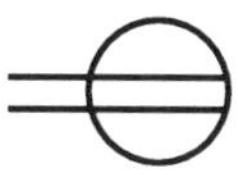
S
switch and receptacle

floor receptacle

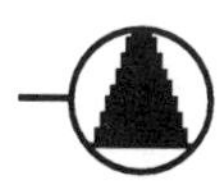
DW
special-purpose dishwasher

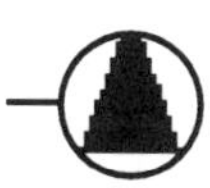
HP
special-purpose heat pump

CD
special-purpose clothes dryer

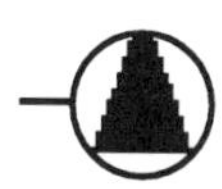
AC
special-purpose air conditioner

PLAN READING

ELECTRICAL SYMBOLS

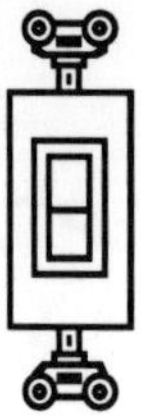

SWITCHED OUTLETS

S	S_2	S_3	S_4	S_D
single-pole switch	double-pole switch	three-way switch	four-way switch	automatic door switch

S_E	S_K	S_P	S_{CB}
electrolier switch	key-operated switch	pilot lamp & switch	circuit breaker

S_{MC}	S_{MC}	S_{WP}	S_{DS}
momentary switch	remote control switch	weatherproof switch	dimmer switch

S_F	S_L	S_{LM}	S_T
fused switch	low-voltage switch	low-voltage master switch	time switch

ELECTRICAL SYMBOLS

AUXILIARY SYSTEMS

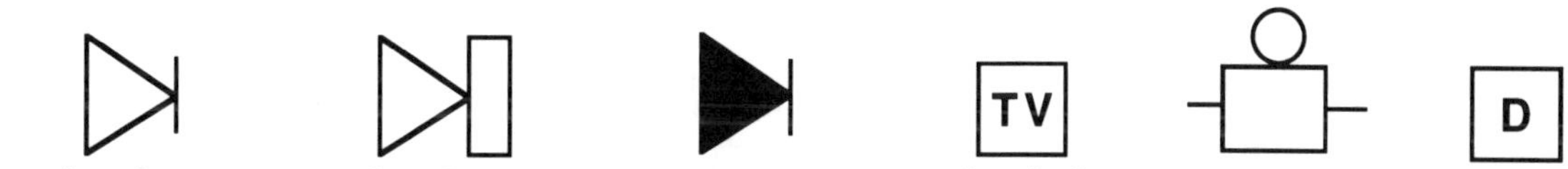

telephone | telephone switchboard | outside telephone | television | bell | electric door opener

chimes | annunciator | horn | push button | bell ringing transformer | buzzer

smoke detector | fire alarm bell | fire alarm horn | fire alarm station | automatic fire alarm station | city fire alarm station

watchman's station | watchman's central station | nurse's signal plug | maid's signal plug | interconnection box

signal central station | radio | battery | thermostat | humidistat | speaker

ABBREVIATIONS

Air Cond.	Air Conditioning
AFF	Above Finish Floor
AL	Aluminum
AC	Alternating current
AIC	Ampere Interrupting Capacity
AIR	Ampere Interrupting Rating
APPROX.	Approximate
ARCH	Architectural
ASPH	Asphalt
AV	Average
AWG	American Wire Gage
BSMT	Basement
B	Bathroom
BR	Bedroom
BP	Blueprint
BLDG	Building
BL	Building Line
C	Capacitor
CAB	Cabinet
CLG	Ceiling
CEL	Cellar
CEM	Cement
CTR	Center
C to C	Center to Center
CIR or CKT	Circuit
CL	Center Line
C.B. or CIR BKR	Circuit Breaker
CO	Clean Out
C, CL or CLO	Closet
CW	Cold Water
COL	Column
COM or CPRSR	Compressor
CONC B	Concrete Block
CND	Conduit
C.O. or -C.O.-	Conduit Only
CONT	Contract
CU or COP	Copper
CU FT.	Cubic Foot
D	Diode or Rectifier
DC	Direct Current
DET	Detail
DIAG	Diagram
DIA	Diameter
DIM	Dimension
DISTR	Distribution
DO	Door Opener
DT	Dust Tight or Double Throw
DW	Dishwasher
DWG	Drawing
D	Dryer
EMT	Electrical Metallic Tubing
E to E	End to End
ENT	Entrance
EOR	End Of Run
EP	Explosion Proof
ESP	Emergency Switch Panel
EXC	Excavate
EXH	Exhaust
EXIST	Existing
EXP JT	Expansion Joint
F	Fan or Fuse
FIN	Finish
FIN. FL	Finish Floor
FP	Fire Place
FPRF	Fireproof
FIX	Fixture
FL	Flashing, Floor or Flush
FLA	Full Load Amps
FD	Floor Drain
FLUOR	Fluorescent
' or FT	Foot
FTG	Footing
FND	Foundation
FPN	Fine Print Note
FRM	Frame
FUR	Future
F.B.O.	Furnished By Others
GAR	Garage
GA	Gauge
GD	Garbage Disposal
GFCI	Ground Fault Interrupter
GRND or GND	Ground
HACR	Heat-AC-Refrigeration
HGT, H or HT	Height
HOA	Hand-Off-Automatic
HID	High Intensity Discharge
HOR	Horizontal
HB	Hose Bibb
HP	Horsepower
HTG	Heating
HW	Hot Water
HWH	Hot Water Heater
HZ	Hertz (cycles per second)
I	Current
" or IN	Inch
ID	Inside Diameter
IMC	Intermediate Conduit
INT	Interior
JB	Junction Box
K	One Thousand (1000)
KCMIL	Thousand Circular Mils
KIT	Kitchen

ABBREVIATIONS

KVA	Kilo Volt Amperes
KW	Kilo Watt
KWH	Kilo Watt Hour
L	Length or Lamps
LTH	Lath
LAU	Laundry
LAV	Lavatory
LG or LGTH	Length
LOA	Length Overall
LOC	Location
LEV	Level
LT	Light
LTG	Lighting
LTS	Lights
LM	lumens
L CL	Linen Closet
LR	Living Room
MAN	Manual
MN	Main
MATL	Material
MAX	Maximum
M	Thousand (1000)
MCC	Motor Control Center
MCM	Thousand Circular Mils
MDP	Main Distribution Panel
MECH	Mechanical
MIN	Minimum
MISC	Miscellaneous
MOR	Middle Of Run
MTR or MOT	Motor
MS	Motor Switch
MT	Empty
MTD	Mounted
N	North
NEU	Neutral
NL	Night Light
NO	Number
NTS	Not To Scale
# or NO.	Number
O.C.B.	Oil Circuit Breaker
OC	Overcurrent
O.C.	On Center
OPNG	Opening
OUT	Outlet
OD	Outside Diameter
OA	Overall
OUT	Outlet
OVHD	Overhead
PB	Push Button
PC	Pull Chain
PF	Power Factor
PL	Pilot Light
PNL	Panel
PTN	Partition
PERP	Perpendicular
PL or PLAS	Plaster
PLAT	Platform
PLBG	Plumbing
P.O.C.	Point Of Connection
P	Porch
PR	Pair
PVC	Polyvinyl Chloride
PREFAB	Prefabricated
PS	Pull Switch
RAD	Radiator
R	Recessed, range or reset
RECEPT or RCPT	Receptacle
RDM	Random
REC	Recessed
REF	Refrigerator
REG	Registered
REINF	Reinforce
RPM	Revolutions per minute
REV	Reverse or Revision
RF	Roof
RM or R	Room
RH	Range Hood
RT	Raintight
SC	Scale
SCH	Schedule
S	Switch, Scuttle or South
SEL	Select
SE or SERV	Service
SEW	Sewer
SH	Sheet or Shower
SK or S	Sink
SP	Spare
SPEC	Specifications
SQ	Square
SQ FT	Square Feet
SQ IN	Square Inch
STD	Standard
STL	Steel
STK	Stock
STG	Storage
STR	Start
SW or S	Switch
T	Toilet
TEL	Telephone
THERMO	Thermostat
THK or T	Thick or Thickness
TRANS or XFMR	Transformer

ABBREVIATIONS

TERM	Terminal
U.O.N.	Unless Otherwise Noted
U.O.S.	Under Other Specifications
UL	Underwriters Laboratory
UNEXC	Unexcavated
UNF	Unfinished
URM	Utility Room
V	Voltage or Vent
VA	Volt Amps
VD	Voltage Drop
VERT	Vertical
VT	Vapor Tight
WM	Washing Machine
W	Watt or Water
WC	Water Closet
WH	Water Heater
WP	Waterproof or Weatherproof
WT	Weight
W	West or Width
WTH	Width
XFMR	Transformer
Z	Reactance

OTHER BUILDING ABBREVIATIONS

AB	Anchor Bolt
ABV	Above
AC	Acoustical
ACFL	Access Floor
ACPL	Acoustical Plaster
ACT	Acoustical Tile
AD	Area Drain
ADD	Addendum
ADH	Adhesive
ADJ	Adjacent
AGG	Aggregate
ALT	Alternate
ANC	Anchor
ANOD	Anodized
AP	Access Panel
APX	Approximate
ARCH	Architect
ASB	Asbestos
ASPH	Asphalt
AUTO	Automatic
BBD	Bulletin Board
BD	Board
BEL	Below
BETW	Between
BIT	Bituminous
BLDG	Building
BLK	Block
BLKG	Blocking
BM	Bench Mark
BOT	Bottom
BPL	Bearing Plate
BRG	Bearing
BRK	Brick
BRZ	Bronze
BS	Both Sides
BSMT	Basement
BUR	Built-up Roof
BVL	Beveled
BW	Both Ways
CAB	Cabinet
CB	Catch Basin
CEM	Cement
CER	Ceramic
CFL	Counterflashing
CUFT	Cubic Foot
CG	Corner Guard
CHAM	Chamfer
CHBD	Chalkboard
CLGHT	Ceiling Height
CI	Cast Iron
CIR	Circle
CJ	Control Joint
CK	Caulk or Caulking
CLG	Ceiling
CLR	Clear or Clearance
CLS	Closure
CM	Centimeter
CMU	Concrete Masonry Unit
COL	Column
COMB	Combination
COMP	Compress
COMPT	Compartment
CONC	Concrete
CONST	Construction
CONT	Continuous
CONTR	Contract or Contractor
CORR	Corridor
CPR	Copper
CPT	Carpet
CRS	Course
CS	Countersink
CSMT	Casement

CT	Ceramic Tile
CTR	Counter
CUYD	Cubic Yard
D	Drain
DEM	Demolish
DEP	Depressed
DF	Drinking Fountain
DH	Double Hung
DIAG	Diagonal
DIAM	Diameter
DIM	Dimension
DIV	Division
DL	Dead Load
DMT	Demountable
DN	Down
DPR	Dampproofing
DR	Door
DRB	Drainboard
DS	Downspout
DT	Drain Tile
DET	Detail
DTL	Detail
DW	Dumbwaiter
DWG	Drawing
E	East
EB	Expansion Bolt
EJ	Expansion Joint
EL	Elevation
ELEV	Elevator
EMER	Emergency
ENC	Enclosure
EQ	Equal
EQP	Equipment
ESC	Escalator
EST	Estimate
EWC	Electric Water Cooler
EXCA	Excavate
EXIST	Existing
EXH	Exhaust
EXP	Exposed
EXT	Exterior
ES	Each Side
FA	Fire Alarm
FB	Face Brick
FBD	Fiberboard
FBO	Furnished By Others
FBRK	Fire Brick
FD	Floor Drain
FE	Fire Extinguisher
FEC	Fire Extinguisher Cabinet
FF	Finished Floor
FBGL	Fiberglass
FHS	Fire Hose Station
FIN	Finished
FCO	Floor Clean Out
FLG	Flashing
FLR	Floor or Flooring
FLEX	Flexible
FN	Fence
FND	Foundation
FOC	Face Of Concrete
FOF	Face Of Finish
FOM	Face Of Masonry
FOS	Face Of Studs
FP	Fireproof
FPL	Fireplace
FR	Frame or Framing
FRC	Fire Retardant
FTG	Footing
FUR	Furred or Furring
GA	Gauge
GB	Grab Bar
GC	General Contractor
GD	Grade
GKT	Gasket
GL	Glass or Glazing
GLB	Glass Block
GLF	Glass Fiber
GPDW	Gypsum Drywall
GRN	Granite
GST	Glazed Structural Tile
GT	Grout
GALV	Galvanized
GVL	Gravel
HB	Hose Bibb
HBD	Hardboard
HC	Hollow Core
HDR	Header
HDW	Hardware
HM	Hollow Metal
HDRZ	Horizontal
HP	High Point
HT	Height
HTG	Heating
HVAC	Heat, Ventilating & AC
HWD	Hardwood
HWH	Hot Water Heater
ID	Inside Diameter
INCIN	Incinerator
INCL	Include
INS	Insulation
INT	Interior
INTER	Intermediate
INV	Invert
JC	Janitor's Closet
JF	Joint Filler
JST	Joist
JT	Joint

PLAN READING

KIT	Kitchen
KO	Knock Out
KPL	Kick Plate
L	Length
LAB	Laboratory
LAD	Ladder
LAM	Laminate
LAV	Lavatory
LB	Lag Bolt
LBL	Label
LH	Left Hand
LL	Live Load
LPT	Low Point
LT	Light
LTL	Lintel
LVR	Louver
LW	Lightweight
M	Meter
MAS	Masonry
MAX	Maximum
MBR	Member
MC	Medicine Cabinet
MECH	Mechanical
MED	Medium
MET	Metal
MFR	Manufacturer
MH	Manhole
MIN	Minimum
MIRR	Mirror
MISC	Miscellaneous
MLD	Moulding
MM	Millimeter
MEM	Membrane
MO	Masonry Opening
MOV	Movable
MR	Mop Receptor
MRB	Marble
MTFR	Metal Furring
MAT	Material
MULL	Mullion
MWK	Millwork
N	North
NAT	Natural
NIC	Not In Contract
NOM	Nominal
NR	Noise Reduction
NRC	Noise Reduction Coefficient
NTS	Not To Scale
OA	Overall
OC	On Center
OD	Outside Diameter
OH	Overhead
OPG	Opening
OPP	Opposite
O/O	Out To Out
PAR	Parallel
PB	Panic Bar
PBD	Particle Board
PCC	Precast Concrete
PCF	Pound Per Cubic Foot
PE	Porcelain Enamel
PED	Pedestal
PERF	Perforated
PFL	Pounds Per Lineal Foot
PREFIN	Prefinished
PK	Parking
PLG	Plate Glass
PL	Property Line or Plate
PLAM	Plastic Laminate
PLAS	Plaster
PNL	Panel
PNT	Paint
PSC	Prestressed Concrete
PSF	Pounds Per Square Foot
PSI	Pounds Per Square Inch
PT	Point or Pressure Treat
PTC	Post Tensioned Concrete
PTD	Paper Towel Receptor
PVMT	Pavement
PLYWD	Plywood
QT	Quarry Tile
R	Riser
RA	Return Air
RAD	Radius
RB	Rubber Base
RBT	Rubber Tile
RCP	Reinforced Concrete Pipe
RD	Roof Drain
REINF	Reinforced
REF	Reference or Refrigerator
REG	Register
REM	Remove
RES	Resilient
RET	Return
REV	Revision
REFL	Reflective
RH	Right Hand
S	South
SC	Solid Core
SCH	Schedule
SCN	Screen
SD	Storm Drain
SEC	Section
SFGL	Safety Glass
SG	Sheet Glass
SHT	Sheet
SHTH	Sheathing
SIM	Similar

SKL	Skylight
SLT	Sealant
SP	Soundproof
SPEC	Specification
SPK	Speaker
SQ	Square
SK	Sink
SST	Stainless Steel
STA	Station
STD	Standard
STG	Seating
STL	Steel
STO	Storage
STR	Structural
SUS	Suspended
SYM	Symmetrical
SYN	Synthetic
SYS	System
T	Tread
TB	Towel Bar
TEL	Telephone
T&G	Tongue & Groove
THK	Thick or Thickness
THR	Threshold
TKDB	Tackboard
TOL	Tolerance
TPD	Toilet Paper Dispenser
TP	Toilct Partition
TR	Transom
TSL	Top Of Slab
TST	Top Of Steel
TV	Television
TYP	Typical
TZ	Terrazzo
UC	Undercut
UNF	Unfinished
UNO	Unless Otherwise Noted
UR	Urinal
Var	Varnish or Variance
VB	Vinyl Base
VB	Vapor Barrier
VCI	Vinyl Composition Tile
VERT	Vertical
VF	Vinyl Fabric
VIN	Vinyl
VJ	V-Joint
VNR	Veneer
VT	Vinyl Tile
VWC	Vinyl Wallcovering
W	West, Width or Wide
WB	Wood Base
WC	Water Closet or Wallcovering
WD	Wood
WG	Wired Glass
WH	Wall Hung
WI	Wrought Iron
WIN	Window
WM	Wire Mesh
W/	With
W/O	Without
WP	Waterproofing
WR	Water Repellent
WS	Water Stop
WSCT	Wainscot
WTW	Wall To Wall
WWF	Welded Wire Fabric

Some drawings include a symbol schedule as shown below. Not all electrical designers use the same symbols as they modify them to their desires. Always check the drawings for a symbol schedule.

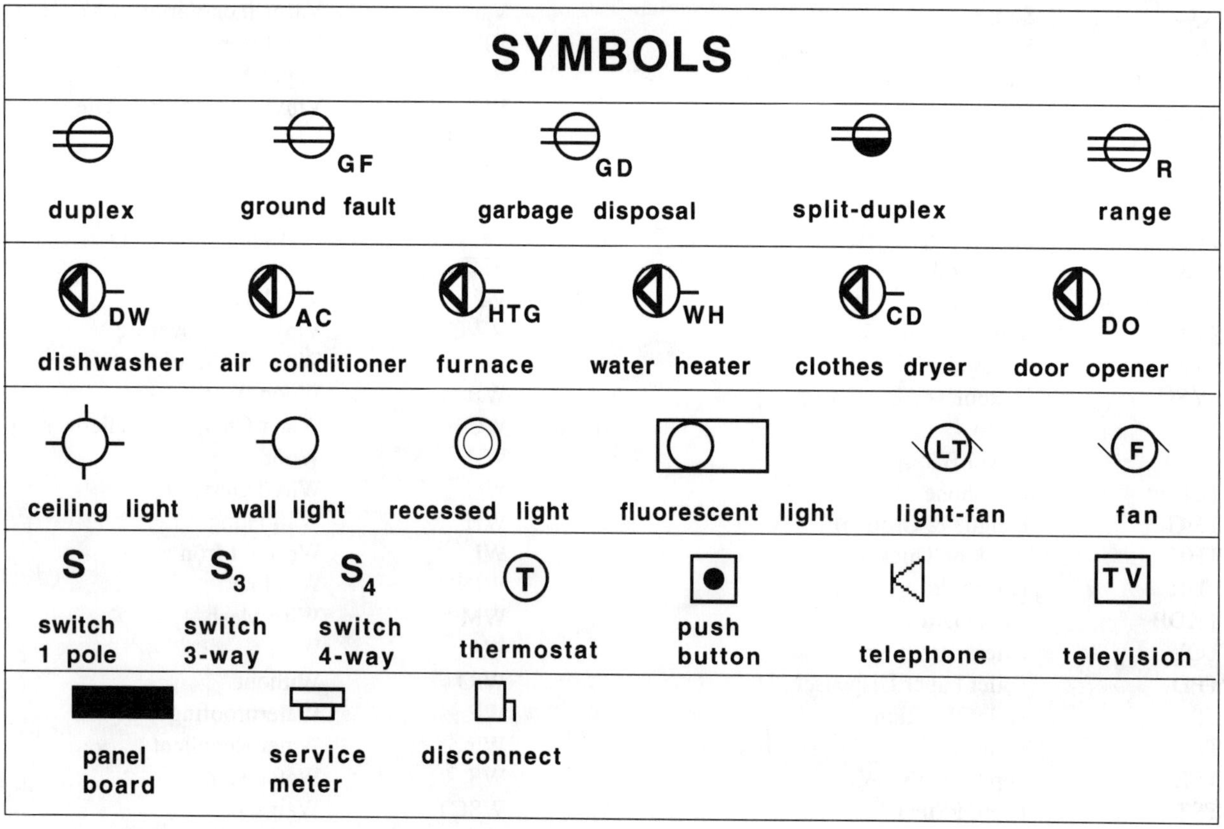

A receptacle schedule is often found on electrical drawings where you have a variety of receptacle types.

RECEPTACLE TYPE SCHEDULE						
Symbol	Amp rating	Wires-Poles	Voltage rating	NEMA type	Configuration	Remarks
	20	3-W 2-P	125	5-20		duplex
R	50	3-W 3-P	125/ 250	10-50		range
CD	30	3-W 3-P	125/ 250	10-30		dryer

The Code states each lighting and appliance panelboard shall be individually protected on the supply side by not more than two main circuit breakers or two sets of fuses having a combined rating not greater than that of the panelboard. The exception to this rule is, individual protection for a lighting and appliance panelboard shall **NOT** be required if the panelboard **FEEDER** has overcurrent protection not greater than the rating of the panelboard.

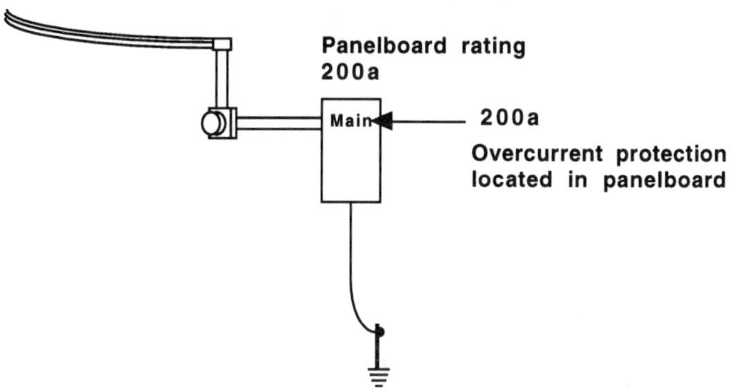

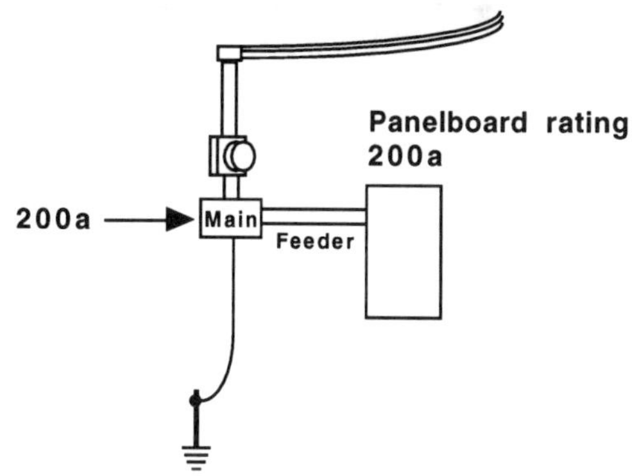

A panelboard might be a service equipment panelboard, distribution panelboard, or a branch-circuit panelboard. A single panelboard could combine two of these functions, or all three within a single cabinet.

If the panelboard is used as a service equipment panelboard, it must be marked **SUITABLE FOR SERVICE EQUIPMENT.**

The most common panelboard is the lighting and appliance panelboard. No more than 42 overcurrent devices (in addition to the main) may be installed.

It is very important to understand the definition of a **LIGHTING AND APPLIANCE** branch circuit panelboard.

The Code defines a lighting and appliance panelboard as one having **MORE** than 10% of its overcurrent devices rated **30 AMPS OR LESS**, for which **NEUTRAL CONNECTIONS** are provided.

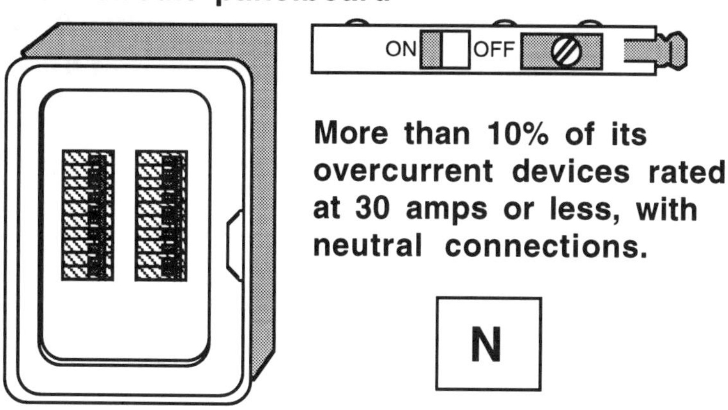

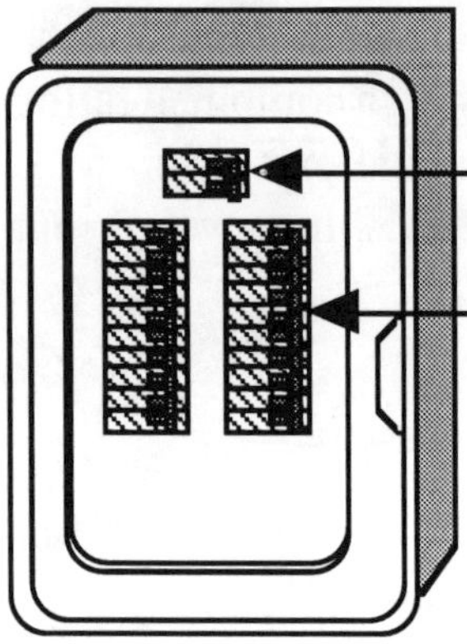

The lighting and appliance panelboard can contain up to 42 overcurrent devices maximum. The main overcurrent device is NOT counted.

1-POLE COUNTS 1

2-POLE COUNTS 2

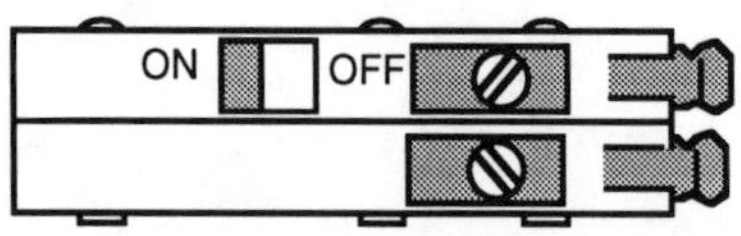

3-POLE COUNTS 3

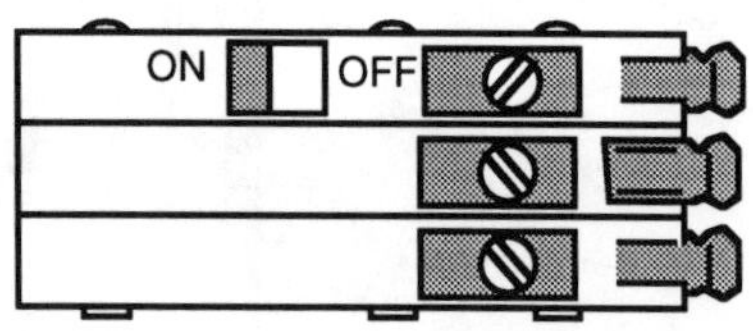

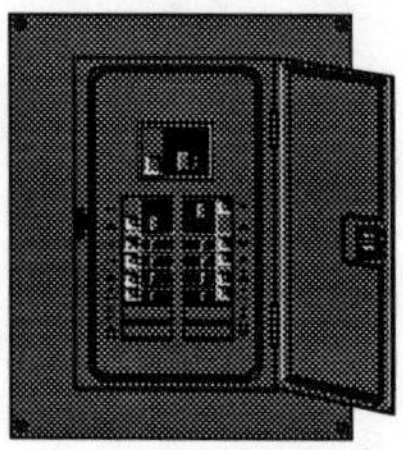

Different manufacturers of panelboards make different sizes of ampere ratings and spaces for single-phase and three-phase panelboards.

Manufacturer "X"

SINGLE-PHASE

Ampere Rating	Spaces
60a	12-18-24
100a	12-18-24
150a	30-36-42
225a	12-18-24-30-36-42

Manufacturer "Y"

SINGLE-PHASE

Ampere Rating	Spaces
100a	12-20
125a	30-42
150a	30-42
175a	30-42
200a	30-42
225a	30-42

Manufacturer "X"

THREE-PHASE

Ampere Rating	Spaces
100a	12-18-24-30-36-42
150a	12-18-24-30-36-42
225a	12-18-24-30-36-42
250a	12-18-24-30-36-42
400a	12-18-24-30-36-42
600a	12-18-24-30-36-42

Manufacturer "X"

THREE-PHASE

Ampere Rating	Spaces
100a	12-24-30-42
110a	12-24-30-42
125a	12-24-30-42
150a	12-24-30-42
175a	12-24-30-42
200a	12-24-30-42
225a	12-24-30-42
250a	12-24-30-42
300a	12-24-30-42
350a	12-24-30-42
400a	12-24-30-42
600a	12-24-30-42

SLAB DRAWING

The following pages show drawings with conduits installed in the concrete slab. The conduits must be secured so they don't move during the pouring of the concrete. The ends of the stub-ups must be capped to prevent concrete from entering.

TYPICAL STUB-UP DETAIL

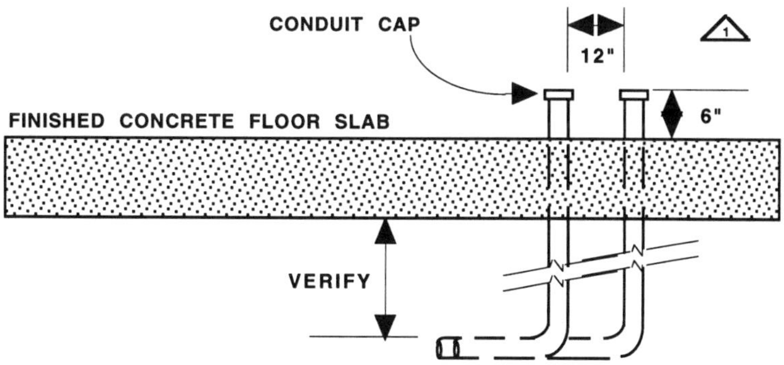

GENERAL NOTES:
1. ALL CONDUIT SHOWN IN DETAILS ARE 3/4" UNLESS OTHERWISE NOTED.

SLAB DRAWING

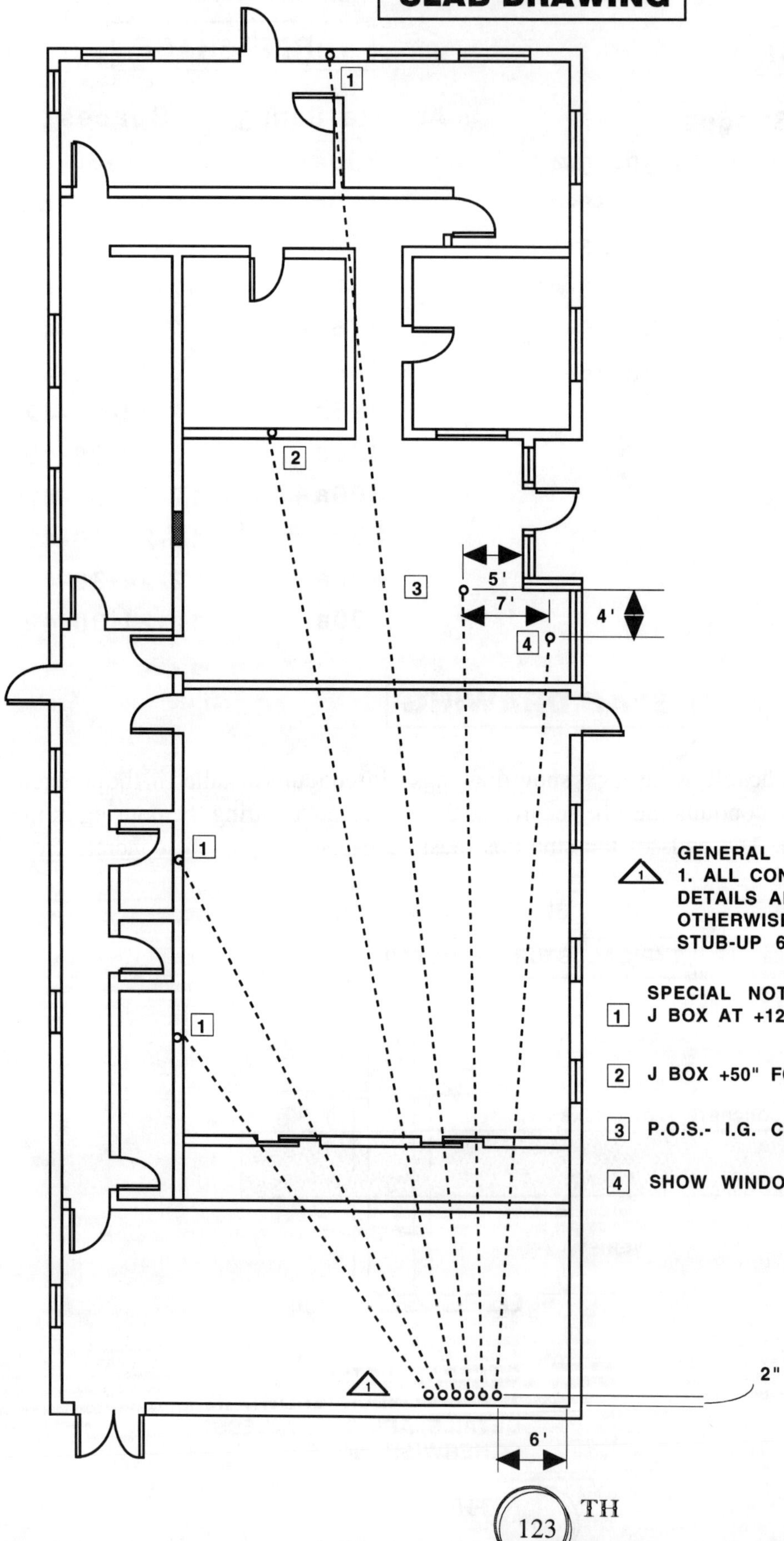

GENERAL NOTES:

1. ALL CONDUIT SHOWN IN DETAILS ARE 3/4" UNLESS OTHERWISE NOTED. STUB-UP 6".

SPECIAL NOTES:

1 J BOX AT +12" FOR RECEPTACLE

2 J BOX +50" FOR COPY MACHINE

3 P.O.S.- I.G. CASH REGISTER

4 SHOW WINDOW LIGHTS

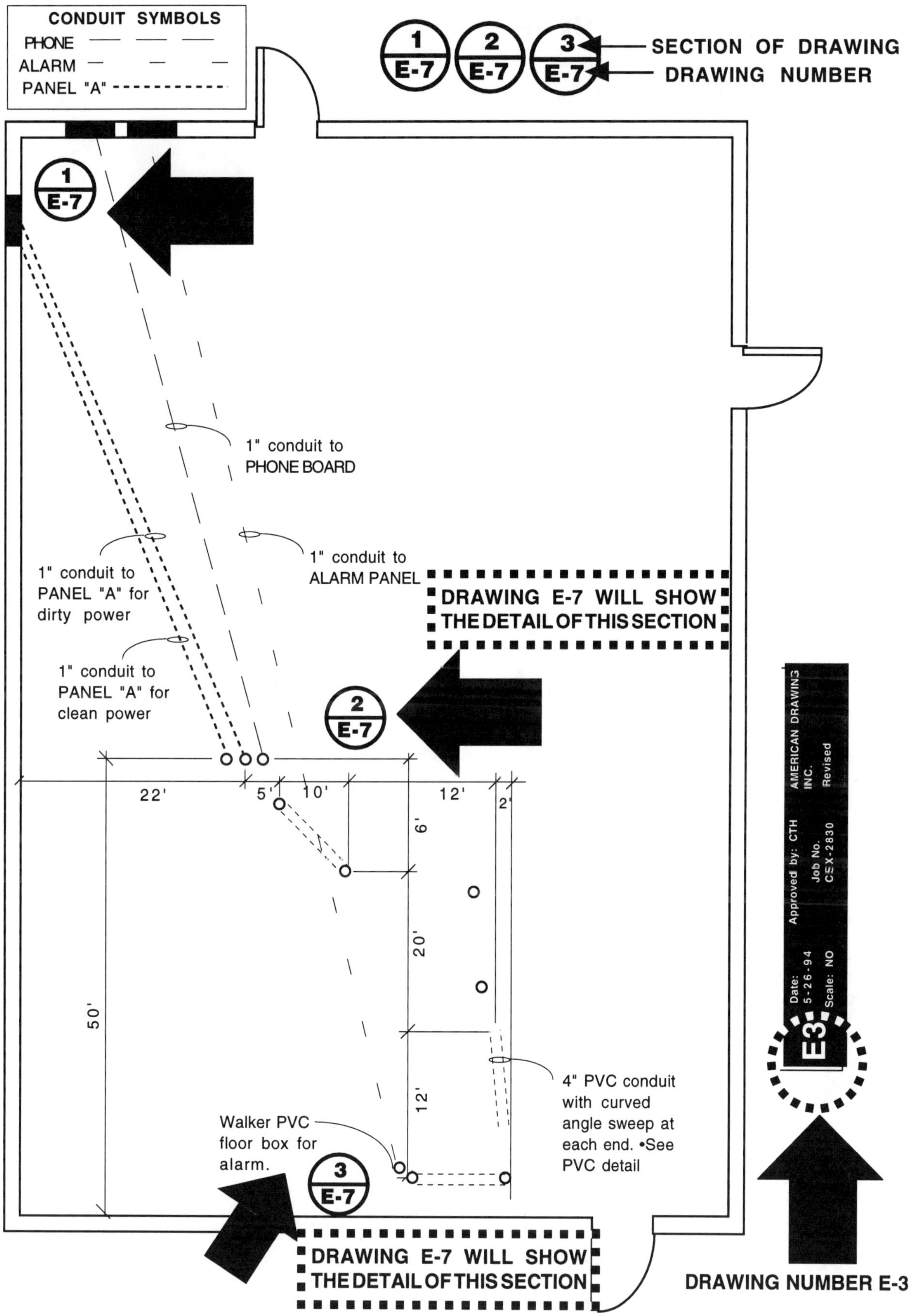
CONDUIT SYMBOLS
PHONE
ALARM
PANEL "A"
SECTION OF DRAWING
DRAWING NUMBER
1 E-7
2 E-7
3 E-7
1" conduit to PHONE BOARD
1" conduit to ALARM PANEL
1" conduit to PANEL "A" for dirty power
1" conduit to PANEL "A" for clean power
DRAWING E-7 WILL SHOW THE DETAIL OF THIS SECTION
22'
5'
10'
12'
2'
6'
20'
12'
50'
4" PVC conduit with curved angle sweep at each end. •See PVC detail
Walker PVC floor box for alarm.
DRAWING E-7 WILL SHOW THE DETAIL OF THIS SECTION
AMERICAN DRAWING INC.
Revised
Approved by: CTH
Job No. CSX-2830
Date: 5-26-94
Scale: NO
E3
DRAWING NUMBER E-3

DETAIL DRAWINGS

Drawing E-7 shows details from drawing E-3.

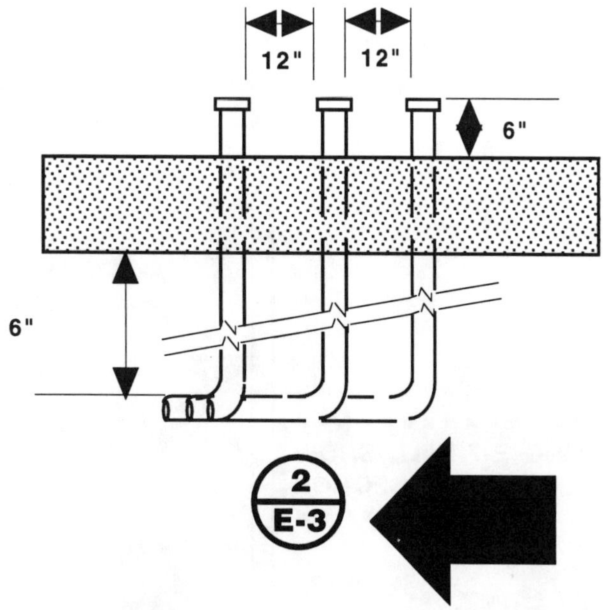

This detail is of SECTION 2
on DRAWING E-3

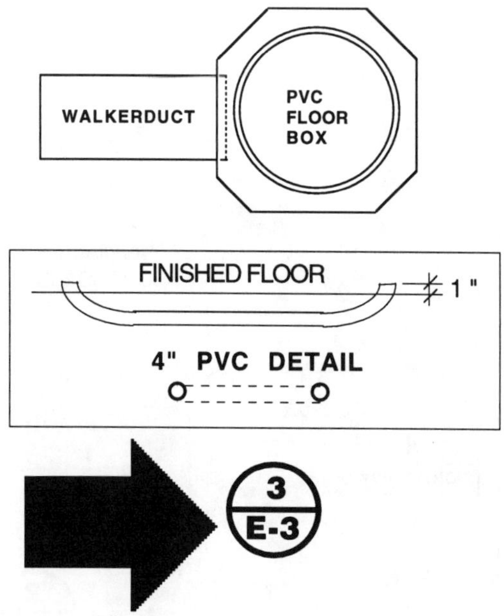

This detail is of SECTION 3
on DRAWING E-3

DRAWING NUMBER E-7

E7	Date: 5-26-94 Scale: NO	Approved by: CTH Job No. CSX-2830	AMERICAN DRAWING INC. Revised

LIGHTING PLAN

The lighting plan shows the layout of the fixtures and the dimensions they are to be installed.

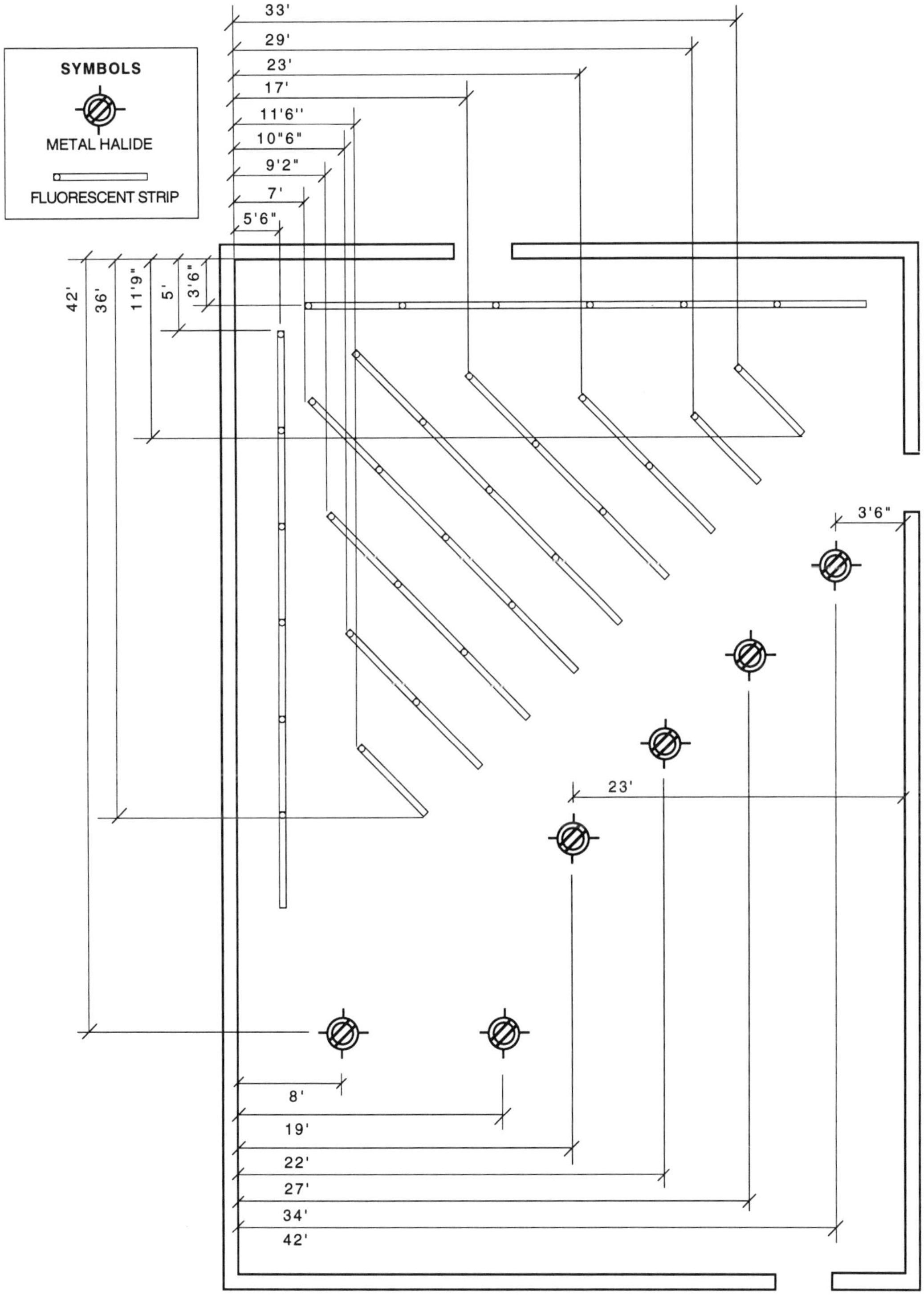

SPECIFICATIONS

Specifications for a construction project is the written description of what is required by the owner, engineer, and architect. Specifications give the grade of materials and the manner in which they are to be installed. Electricians don't always have access to the specifications. In this case they follow the schedule which will list the type of equipment to be installed, including notes.

1. Work Included:

A. Conduit, Wires and Cable
B. Boxes and Wiring devices
C. Safety Switches
D. Service Entrance and Utility Coordination
E. Data Cable (Material furnished by Owner)
F. Conduit for Thermostats (Wiring by Mechanical Contractor)
G. Conduit for Telephone System
H. Conduit for Security System
I. Field Testing
J. Utility Company fees and charges for service to the project, including setting the meters
•Note: Refundable deposits for permanent service will be paid by the owner.

2. Quality Requirements:

A. Provide UL approved materials and devices
B. Comply with NFPA 70 National Electrical Code as minimum standards and local Codes

3. Conduit:

A. Rigid:	Galvanized steel with threaded couplings. Intermediate grade may be used instead of rigid where permitted by the Code.
B. Thin Wall (EMT):	Galvanized steel with liquid tight compression type fittings in concrete slabs and set screw fittings in dry locations.
C. Nonmetallic:	Rigid PVC, Schedule 40, with factory made bends, joints solvent welded.
D. Flexible Conduit:	Spirally wound interlocked steel armored raceway conforming to NEC 350 and UL.

4. Wires and Cables:

A. Soft drawn copper of not less than 98% conductivity.
B. Sizes #8 and larger shall be stranded.
C. Minimum size wire permittcd #12. Use #10 if circuit exceeds 100 feet and #8 where noted.
D. Use 600v THW insulation on building and branch circuit wiring unless otherwise indicated on the drawings. THHN insulation may be used in dry locations only.
E. Data cable size #24 AWG, 3 pair, provided by owner upon notice by contractor of required delivery date.

5. Outlet, Switch, Junction and Pull Boxes:

A. Boxes: Galvanized steel conforming to NEC and UL Requirements
B. Exterior Boxes: Weatherproof type threaded hubs and gasket covers

6. Wiring Devices:

A. Wall Switches: Specification grade, quiet slow make-slow break design, toggle handle, rated 15 amps, 120v, color brown
B. Standard Receptacles: Specification grade, full gang size, polarized duplex, parallel blade, U-grounding slot, rated 20 amps, 120v, color brown
C. Device Plates: General use, Bakelite with smooth finish and rolled outer edges, color brown
D. Safety Switches: General Duty, fusible type with NEMA Type 1 enclosure indoors, and type 3R enclosure outdoors

6. Conduit Installation:

A. Use rigid conduit in earth and where exposed to weather. Make joints with standard thread couplings or unions. Protect newly cut conduit threads with graphite grease or other rust resistive, non-insulating compound. Use double locknuts at terminations.
B. Use EMT only in dry, protected areas where not subject to mechanical damage.
C. PVC conduit may be used for underground feeders from service entrance to panels, to site lighting, and to pylon/billboard signs only. Use rigid steel, long radius elbows on risers from underground conduit. Provide "THW" copper grounding conductors for entire length of nonmetallic conduit and connect to rigid steel.
D. Provide moisture seal where conduits penetrate outside walls or slabs-on-grade.
E. Secure conduits to building structure at intervals of not more than 5 feet for sizes up to and including, 2 inch; and at intervals not to exceed 8 feet for larger sizes.
F. Support conduit in concrete slab construction at every 5 feet or less and secure with #12 AWG wire.
G. Run conduits concealed. Exposed runs, where approved by the architect, shall be installed in neat, symmetrical lines parallel to, or at right angles to, building lines. Tight to metal deck.
H. Install conduit 6 inch minimum from hot water lines.
I. Terminate conduit at equipment subject to vibration with flexible connections. In wet locations, use Neoprene or Polychloroprene covered flex with water tight fittings. Provide for continuity of ground.
J. Provide conduit system for mechanical control wiring.
K. Provide conduit system from telephone system from terminal panelboard to counter location, as noted on plans.
L. Provide conduit for telephone service entrance. Entrance shall be underground from telephone equipment pedestal, and run to phone equipment board.

7. Conductor Installation:

A. Maintain color coding throughout the electric system.
B. Where the distance between the panelboard and the first outlet in a branch circuit is more than 60 feet, use #10 AWG wire or larger.
C. Make conductor length for parallel feeders identical.
D. Provide solderless connections only.

8. Outlet, Switch and Junction Box Installation:

A. Install flush type boxes slightly recessed to allow full perimeter contact of device plates.
B. Install boxes for convenience outlets 12 inches above floor, unless otherwise indicated on the drawings.
C. Install switch boxes 4' 6" above the floor, unless otherwise indicated on the drawings.
D. In masonry construction, the height of boxes may be adjusted slightly to fit coursing.
E. Set boxes for floor outlets flush to floor, unless otherwise indicated on the drawings.
F. "J" Boxes are to be concealed behind the sheet rock.

9. Wiring Device Installation:

A. Place wall switches on strike side of doors unless otherwise indicated on the drawings.
B. Install device plates with perimeter edges in full contact with finish surfaces.

10. Field Testing:

A. Test all circuits with a "Multi" Tester to determine that the system is free of short circuits and that the phase conductors are not grounded.
B. Check motor controllers for proper operation.
C. Check electrical equipment for proper operation.
D. Immediately replace defective or damaged materials and equipment.

11. Data Cable Installation:

A. Cable shall be supplied by owner and received by contractor.
B. Cable lines shall be run straight and parallel, tied together with nonconductive material.
C. Provide 6 feet of extra cable length at each equipment location.
D. Tag each end of Data Cable for each circuit.

OVERHEAD ELECTRICAL SERVICE

1. Components:

A. Conductors:	As required by local utility.
B. Service Rack:	Three (3) or four (4) wire racks, as determined by local utility for specified service.
C. Service Weatherhead:	Cast metal conforming to local utility weatherhead requirements.
D. Service Mast:	Rigid Tubular Metal Fabrication with provisions for attachment of service rack and weatherhead.
E. Meter Socket:	Conform to local utility requirements.
F. Conduit:	Rigid metal conduit complete with watertight fittings.

2. Installation:

A. Install service rack and weatherhead at proper height. Provide mast kit, if required.

B. Extend mast head to provide a minimum ground clearance of 20 feet when overhead service is near dumpster location.

C. Install rigid conduit to weather proof meter socket.

D. Provide terminations, coordinate with local utility company.

3. Panelboards:

A. Metal cabinets with gray enamel finish complete with panel trim having concealed hinges and flush lock, concealed trip clamps and screw cover front; Galvanized steel back with knockouts. Provide two keys for each panel.

B. Factory assembled, three (3) phase - four (4) Wire; and Single Phase - three (3) wire. Full capacity, solid neutral, with sequence style bussing.

4. Circuit Breakers:

A. Molded case, automatic air circuit breakers as supplied with thermal and magnet trip and trip-free position separate from either "ON" or "OFF" positions; interrupting rating of 10,000 AIC.

B. Provide switching duty circuit breakers for branch circuits.

5. Identification of Panels:

A. Provide a laminated plastic plate at the top of panel engraved with designation identical on panelboard schedule or riser diagram. Identify each overcurrent device with a laminated plastic plate engraved with a description of item protected.

B. Provide a typewritten directory to the inside cover indicating loads controlled by each circuit, identical to panel schedules.

LIGHTING

A. Products: Refer to lighting fixture schedule on drawing for types. Provide necessary trim, brackets, and accessories for complete installation. No substitutes will be accepted, unless Architect considers to be equal to what was specified.

B. Lamps: First line quality.

C. Ballasts: Fluorescent - UL listed, Type "P" with automatic resetting thermal protection, rapid start. Metal Halide - UL listed, Type CWA, peak load automatic transformer.

D. Time Clocks: 24 hour dial with two (2) trippers.

E. Control Relay: Six to twelve poles, as required provide additional relays, if required to accommodate site lighting. Balance load between relays.

F. Building: Support light fixtures from structure only. Provide gaskets or necessary accessories to prevent light leaks around trim. Install fixtures in a manner to properly dissipate generated heat. Equip each fixture with owner furnished lamps, as scheduled.

G. Site Installation: Set poles plumb, rigidly in place in accordance with manufacturer's recommendations. Provide necessary brackets, caps and accessories, factory energized light and sign circuits and set time (Dusk). Time clock shall de-energize light and sign circuits at set time (12:00 A.M.) to turn lights off.

Provide necessary exit and emergency egress lighting, clarifying the exit lights will be illuminated to a minimum of five footcandles, and egress lighting will provide not less than one footcandle at the most remote lighted area. For the means of egress.

WIRING DEVICES

STANDARD RECEPTACLE CONFIGURATIONS

NEMA *ANSI*	Receptacle Configuration	Rating	
5-15 *C73.11*		15A 125V	2 POLE 3 WIRE
5-20 *C73.12*		20A 125V	2 POLE 3 WIRE
5-30 *C73.45*		30A 125V	2 POLE 3 WIRE
5-50 *C73.46*		50A 125V	2 POLE 3 WIRE
6-15 *C73.20*		15A 250V	2 POLE 3 WIRE
6-20 *C73.51*		20A 250V	2 POLE 3 WIRE
6-30 *C73.52*		30A 250V	2 POLE 3 WIRE
6-50 *C73.53*		50A 250V	2 POLE 3 WIRE
7-15 *C73.28*		15A 277V	2 POLE 3 WIRE
7-20 *C73.63*		20A 277V	2 POLE 3 WIRE
7-30 *C73.64*		30A 277V	2 POLE 3 WIRE
7-50 *C73.65*		50A 277V	2 POLE 3 WIRE
10-20 *C73.23*		20A 125/250V	3 POLE 3 WIRE
10-30 *C73.24*		30A 125/250V	3 POLE 3 WIRE
10-50 *C73.25*		50A 125/250V	3 POLE 3 WIRE
11-15 *C73.54*		15A 3ϕ 250V	3 POLE 3 WIRE
11-20 *C73.55*		20A 3ϕ 250V	3 POLE 3 WIRE

	Rating	Receptacle Configuration	NEMA *ANSI*
3P3W	30A 3ϕ 250V		11-30 *C73.56*
3P3W	50A 3ϕ 250V		11-50 *C73.57*
3 POLE 4 WIRE	15A 125/250V		14-15 *C73.49*
3 POLE 4 WIRE	20A 125/250V		14-20 *C73.50*
3 POLE 4 WIRE	30A 125/250V		14-30 *C73.16*
3 POLE 4 WIRE	50A 125/250V		14-50 *C73.17*
3 POLE 4 WIRE	60A 125/250V		14-60 *C73.18*
3 POLE 4 WIRE	15A 3ϕ 250V		15-15 *C73.58*
3 POLE 4 WIRE	20A 3ϕ 250V		15-20 *C73.59*
3 POLE 4 WIRE	30A 3ϕ 250V		15-30 *C73.60*
3 POLE 4 WIRE	50A 3ϕ 250V		15-50 *C73.61*
3 POLE 4 WIRE	60A 3ϕ 250V		15-60 *C73.62*
4 POLE 4 WIRE	15A 3ϕY 120/208V		18-15 *C73.15*
4 POLE 4 WIRE	20A 3ϕY 120/208V		18-20 *C73.26*
4 POLE 4 WIRE	30A 3ϕY 120/208V		18-30 *C73.47*
4 POLE 4 WIRE	50A 3ϕY 120/208V		18-50 *C73.48*
4 POLE 4 WIRE	60A 3ϕY 120/208V		18-60 *C73.27*

WIRING DEVICES

STANDARD RECEPTACLE CONFIGURATIONS

NEMA *ANSI*	Receptacle Configuration	Rating	
ML2 *C73.44*		15A 125V	2 POLE 3 WIRE
L5-15 *C73.42*		15A 125V	2 POLE 3 WIRE
L5-20 *C73.72*		20A 125V	2 POLE 3 WIRE
L5-30 *C73.73*		30A 125V	2 POLE 3 WIRE
L6-15 *C73.74*		15A 250V	2 POLE 3 WIRE
L6-20 *C73.75*		20A 250V	2 POLE 3 WIRE
L6-30 *C73.76*		30A 250V	2 POLE 3 WIRE
L7-15 *C73.43*		15A 277V	2 POLE 3 WIRE
L7-20 *C73.77*		20A 277V	2 POLE 3 WIRE
L7-30 *C73.78*		30A 277V	2 POLE 3 WIRE
L8-20 *C73.79*		20A 480V	2 POLE 3 WIRE
L8-30 *C73.80*		30A 480V	2 POLE 3 WIRE
L9-20 *C73.81*		20A 600V	2 POLE 3 WIRE
L9-30 *C73.82*		30A 600V	2 POLE 3 WIRE
ML3 *C73.30*		15A 125/250V	3 POLE 3 WIRE
L10-20 *C73.96*		20A 125/250V	3 POLE 3 WIRE
L10-30 *C73.97*		30A 125/250V	3 POLE 3 WIRE
L11-15 *C73.98*		15A 3φ 250V	3 POLE 3 WIRE
L11-20 *C73.99*		20A 3φ 250V	3 POLE 3 WIRE
L11-30 *C73-100*		30A 3φ 250V	3 POLE 3 WIRE
L12-20 *C73.101*		20A 3φ 480V	3 POLE 3 WIRE

	Rating	Receptacle Configuration	NEMA *ANSI*
3P3W	30A 3φ 480V		L12-30 *C73.102*
3P3W	30A 3φ 600V		L13-30 *C73.103*
3 POLE 4 WIRE	20A 125/250V		L14-20 *C73.83*
3 POLE 4 WIRE	30A 125/250V		L14-30 *C73.84*
3 POLE 4 WIRE	20A 3φ 250V		L15-20 *C73.85*
3 POLE 4 WIRE	30A 3φ 250V		L15-30 *C73.86*
3 POLE 4 WIRE	20A 3φ 480V		L16-20 *C73.87*
3 POLE 4 WIRE	30A 3φ 480V		L16-30 *C73.88*
3 POLE 4 WIRE	30A 3φ 600V		L17-30 *C73.89*
4 POLE 4 WIRE	20A 3φY 120/208V		L18-20 *C73.104*
4 POLE 4 WIRE	30A 3φY 120/208V		L18-30 *C73.105*
4 POLE 4 WIRE	20A 3φY 277/480V		L19-20 *C73.106*
4 POLE 4 WIRE	30A 3φY 277/480V		L19-30 *C73.107*
4 POLE 4 WIRE	20A 3φY 347/600V		L20-20 *C73.108*
4 POLE 4 WIRE	30A 3φY 347/600V		L20-30 *C73.109*
4 POLE 5 WIRE	20A 3φY 120/208V		L21-20 *C73.90*
4 POLE 5 WIRE	30A 3φY 120/208V		L21-30 *C73.91*
4 POLE 5 WIRE	20A 3φY 277/480V		L22-20 *C73.92*
4 POLE 5 WIRE	30A 3φY 277/480V		L22-30 *C73.93*
4 POLE 5 WIRE	20A 3φY 347/600V		L23-20 *C73.94*
4 POLE 5 WIRE	30A 3φY 347/600V		L23-30 *C73.95*

MATERIAL DESIGNATIONS

- EARTH / COMPACTED
- POROUS FILL (STONE OR GRAVEL, ETC.)
- ROCK
- CONCRETE / LIGHTWEIGHT (CONCRETE FILL)
- CONCRETE / STRUCTURAL (CAST-IN-PLACE, PRECAST)
- BRICK / COMMON
- BRICK / FACE
- CONCRETE MASONRY UNITS (SMALL SCALE)
- CONCRETE MASONRY UNITS (LARGE SCALE)
- CAST STONE CONCRETE
- STONE / CUT
- MARBLE
- STONE / RUBBLE
- CLAY TILE / STRUCTURAL
- CLAY TILE / UNGLAZED
- CLAY TILE / GLAZED
- METAL (SMALL SCALE)
- PLYWOOD (LARGE SCALE)
- PLYWOOD (SMALL SCALE)
- WOOD / FINISH
- WOOD / ROUGH
- WOOD / BLOCKING
- INSULATION / LOOSE OR BATT
- INSULATION / RIGID
- GLASS (LARGE SCALE)
- GLASS (SMALL SCALE)
- ACOUSTICAL TILE
- CERAMIC TILE
- GYPSUM WALL BOARD
- PLASTER, SAND, CEMENT, GROUT
- RESILIENT FLOORING
- TERRAZZO

ELEVATION

- CONCRETE, PLASTER, STUCCO (AS INDICATED)
- CONCRETE MASONRY UNITS
- GLAZING
- BRICK

PARTITION CONSTRUCTION PLAN

- WOOD STUCCO
- STEEL STUD
- CMU
- CMU W/FILLED CELLS

CHAPTER 5

EXAM

PLAN READING EXAM 1

1. How many telephones are shown?

(a) 1 (b) 2 (c) 3 (d) 4

2. How many duplex receptacles are shown?

(a) 12 (b) 13 (c) 14 (d) 15

3. Which circuit has lights and a receptacle?

(a) A-14 (b) B-8, 10,12 (c) A-19 (d) A-11

PLAN R-3344

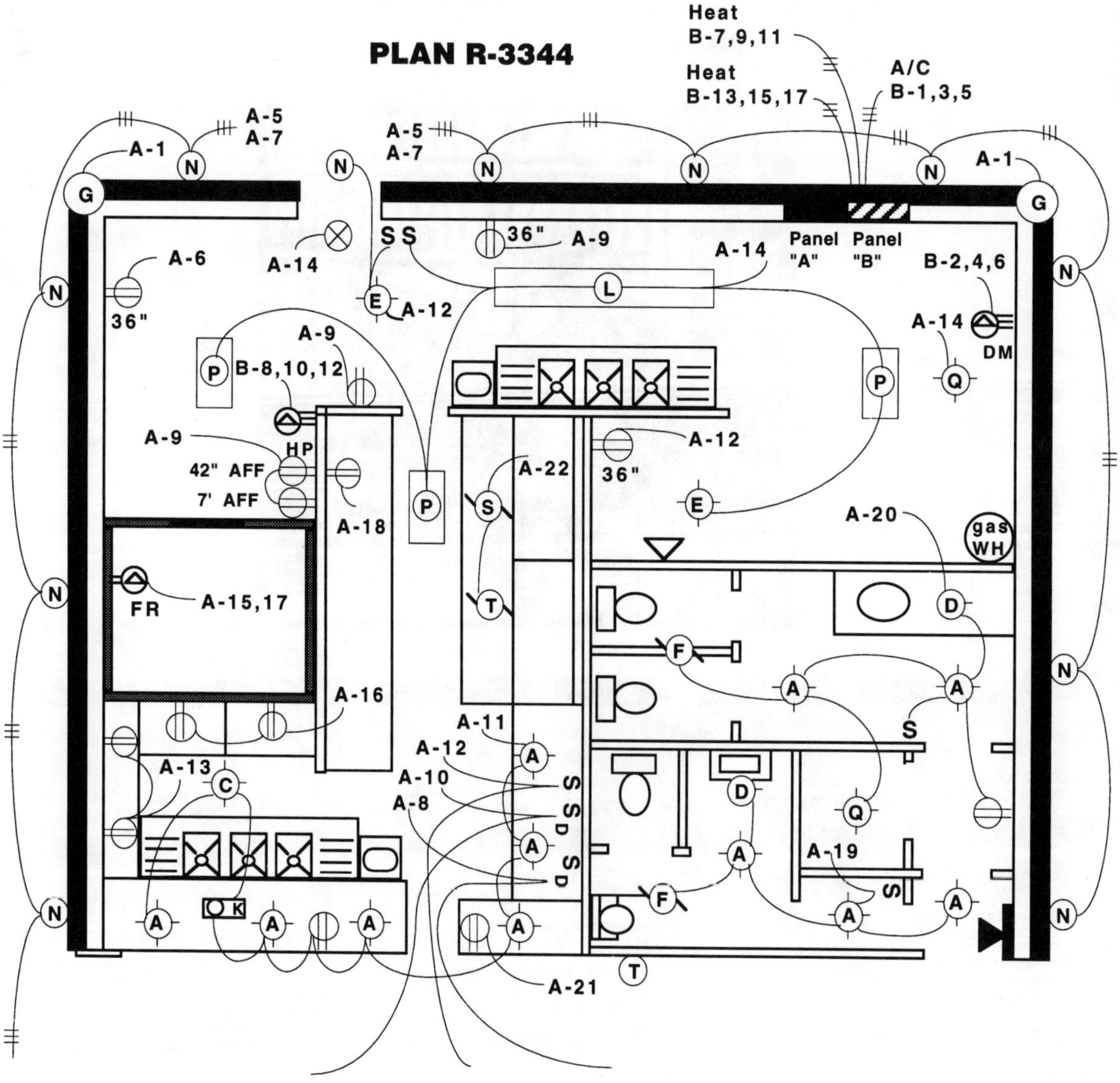

PLAN READING EXAM 2

1. The number of exit lights shown on the drawing "Part B" is ____.

(a) 1 (b) 2 (c) 3 (d) 4

2. The number of interior wall mounted light fixtures on drawing "Part B" is ____.

(a) 22 (b) 18 (c) 15 (d) 30

PART "B" OF PLAN R-3344

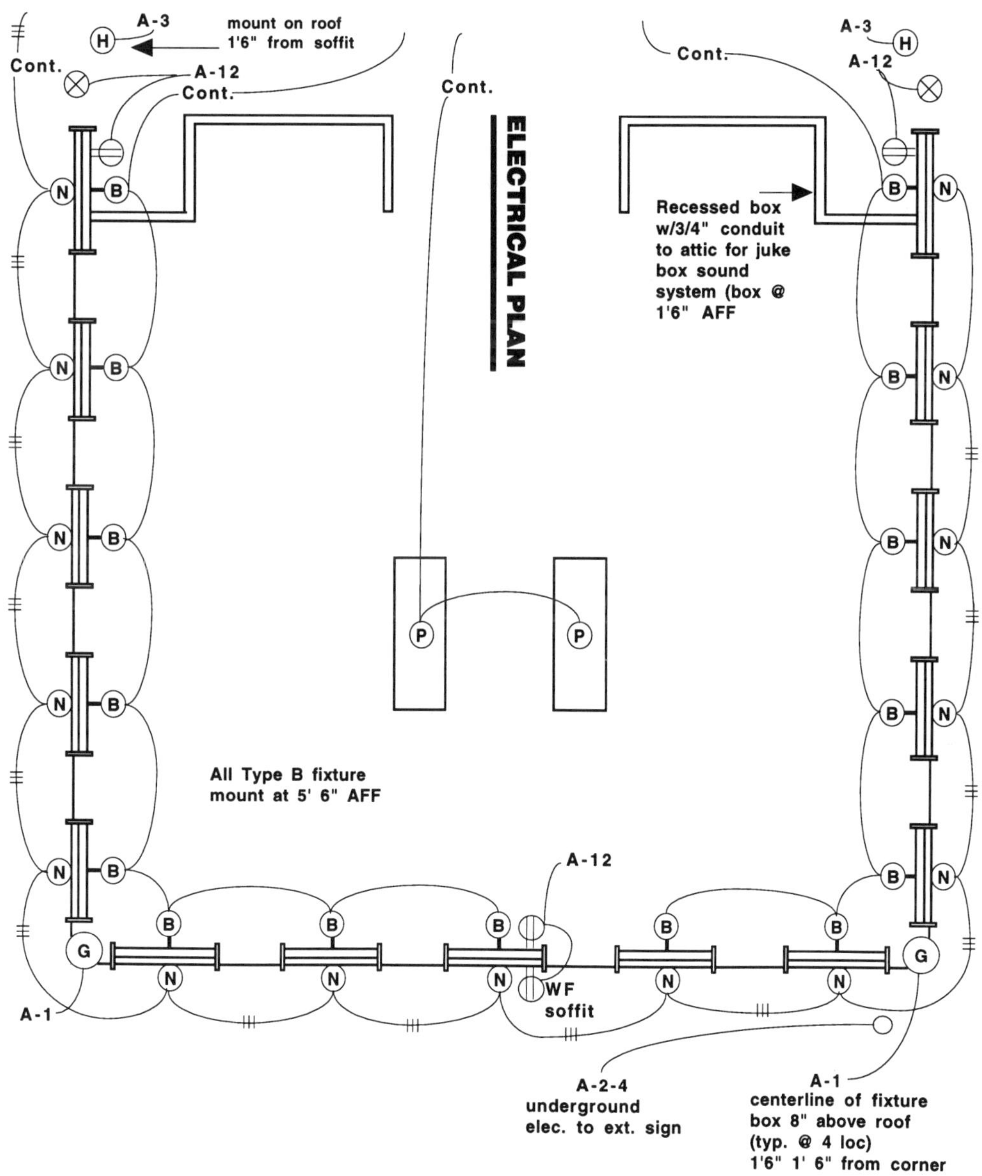

PLAN READING EXAM 3

•Questions on this exam are from the plans of the restaurant **R-3344**.

1. Circuit A-11 has ____ lights.

(a) 7 (b) 8 (c) 9 (d) 6

2. How many "Q" light fixtures are shown?

(a) 1 (b) 2 (c) 3 (d) 4

3. How many 3-way switches are shown?

(a) 0 (b) 1 (c) 2 (d) 4

4. How many "N" light fixtures are shown?

(a) 23 (b) 24 (c) 25 (d) 26

5. How many flourescent fixtures are shown?

(a) 7 (b) 6 (c) 5 (d) 4

6. Circuit A-19 supplies ____ lights.

(a) 5 (b) 4 (c) 3 (d) 2

7. How many lights are connected to circuit A-12?

(a) 2 (b) 3 (c) 5 (d) 6

PLAN READING EXAM 4

•Questions on this exam are from the plans of the restaurant **R-3344**.

1. (T) is the symbol for ____.

(a) timer (b) thermostat (c) fan (d) clock

2. Circuit A-20 supplies ____ receptacle(s).

(a) 1 (b) 2 (c) 3 (d) 4

3. How many duplex receptacles are connected to circuit A-12?

(a) 0 (b) 2 (c) 4 (d) 5

4. Circuit A-1 supplies ____ light(s).

(a) 4 (b) 3 (c) 2 (d) 1

5. How many special receptacles are shown on drawing R-3344?

(a) 4 (b) 3 (c) 2 (d) 1

6. Circuit A-14 supplies ____ lights.

(a) 5 (b) 6 (c) 7 (d) 8

7. Circuit A-9 supplies ____ receptacle(s).

(a) 4 (b) 3 (c) 2 (d) 1

CHAPTER 6

FAULT CURRENTS

FAULT CURRENTS

In recent years utility companies have increased their generating capacity to meet the demand. This also increases the **AVAILABLE SHORT-CIRCUIT CURRENT**.

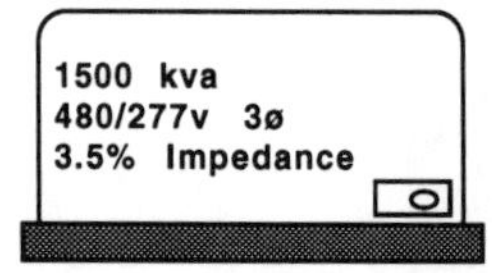

The trend in equipment is to lower impedance for better equipment utilization. Thus, increased capacity and **REDUCED IMPEDANCE** have greatly increased the short-circuit current available today, compared with 10 years ago.

It is important for the inspector to check the plans for the available short circuit current and during inspection make sure the correct device are installed per the approved plans.

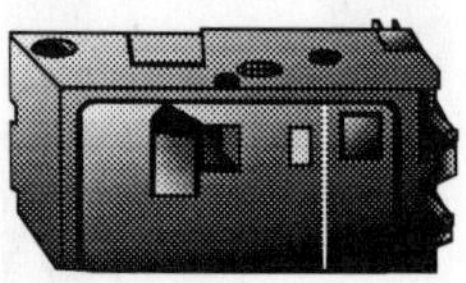

A circuit breaker's interrupting rating can vary with application voltage. The breaker may have an interrupting rating of 18,000 amperes at 240 volts but only 14,000 amperes at 480 volts.

The breaker or fuse must have an interrupting rating equal to or greater than the available short-circuit current at their **LINE SIDE**.

The interrupting rating of the typical breaker used in a residence is 10,000 amperes (•A breaker's interrupting capacity will typically be lower than its interrupting rating.) Larger breakers may have interrupting ratings of 14,000 amperes or higher. Most modern current-limiting fuses have an interrupting rating of 200,000 amperes.

Where the available fault current exceeds the breaker interrupting rating a **CURRENT-LIMITING FUSE** can be used to limit the available short-circuit current to a value the breaker can withstand, thus protecting the breaker. Also the voltage rating must be equal to or greater than the system voltage.

The rating which defines the capacity of a fuse or breaker to maintain its integrity when reacting to fault currents is termed its **INTERRUPTING RATING**.

A fuse or breaker must be able to withstand the destructive energy of short-circuit currents. If a fault current exceeds a level beyond the capability of the fuse or breaker, it may actually rupture, causing additional damage. Thus it is important when applying a fuse or breaker to use one which can sustain the largest potential short-circuit currents.

Modern fuses have an extremely high interrupting rating, can withstand very high fault currents without rupturing.

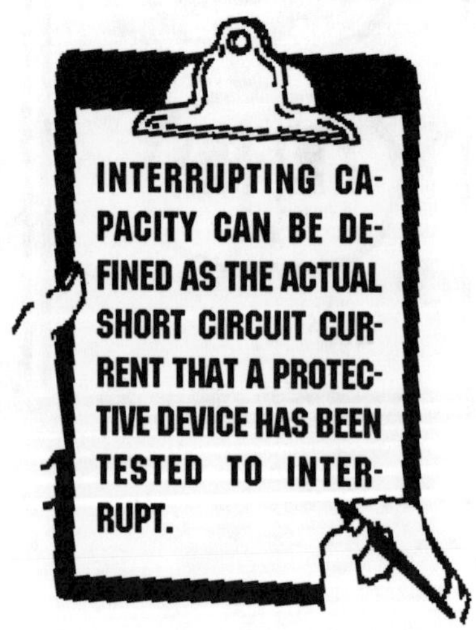

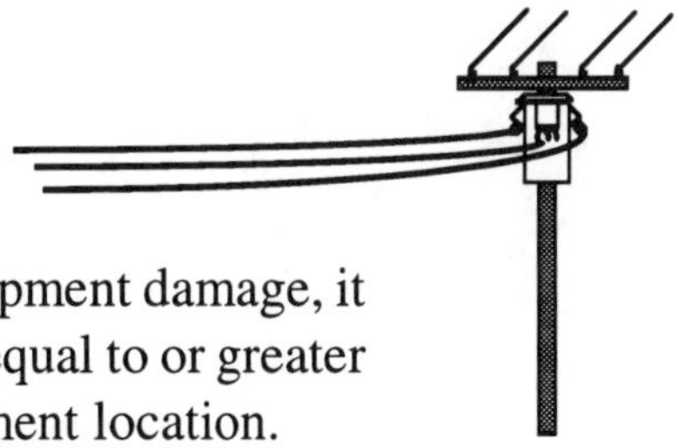

The utility company will advise the **MAXIMUM AVAILABLE SHORT-CIRCUIT AMPERES** available at the user's service connection.

In order to provide for personal safety and to minimize equipment damage, it is absolutely essential to use equipment with **SHORT-CIRCUIT RATINGS** equal to or greater than the available short-circuit current that can occur at the equipment location.

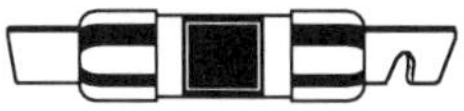

There is one instance when equipment can be applied at a location where the available short-circuit current is higher than the short-circuit rating of the device. This arrangement utilizes a **CURRENT LIMITING FUSE**.

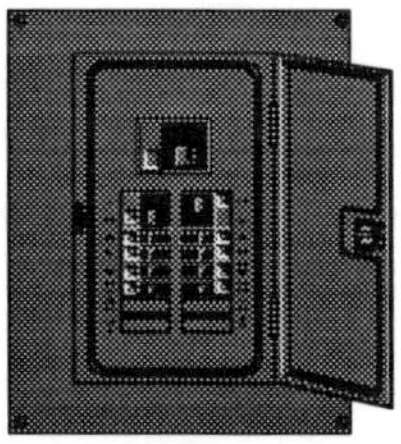

A panelboard containing breakers with a short-circuit rating of 14,000 amperes located where the available short-circuit current is 45,000 amperes is obviously a violation as these breakers do not have an adequate short-circuit rating.

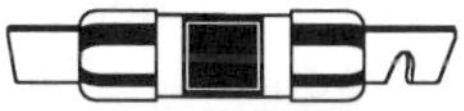

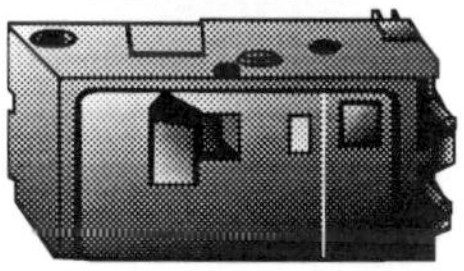

If current limiting fuses are located in the mains of the panelboard or at an upstream location feeding the panelboard, the breakers can be adequately protected.

If a protective device cuts off a short-circuit current in less than 1/2 cycle, before it reaches its total available current, the device is a **CURRENT LIMITING** device. They minimize the need of other components to have high short-circuit current **WITHSTAND** ratings. If not limited, short-circuit currents can reach levels of 40,000 amperes or more in the first 1/2 cycle (.008 seconds) after the start of a short-circuit. The heat produced by short-circuit currents can cause severe insulation damage or even an explosion.

Breakers are mechanical overcurrent protective devices which generally have low interrupting ratings and usually are not current-limiting.

To adequately protect components in modern systems, it is imperative that the overcurrent device be current-limiting.

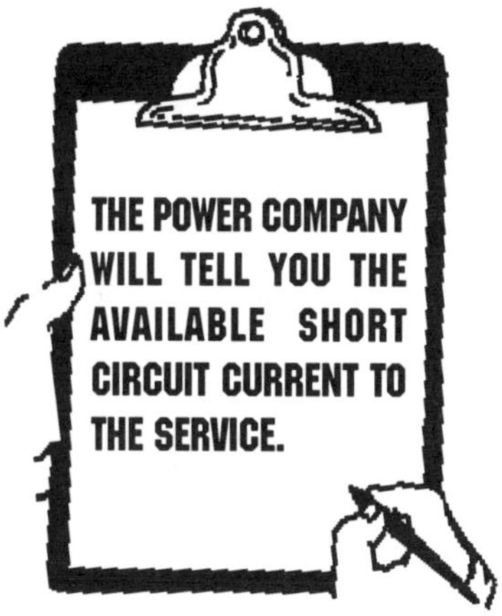

The selection of overcurrent protective devices must include careful designing of the ability of the devices to operate properly and safely on ground faults and short-circuits.

Normally a circuit draws current in proportion to the voltage applied and the impedance of the load. When a short circuit occurs, the source voltage no longer encounters the opposition to the current flow that the normal load had presented. The load is *bypassed* and large currents are impressed upon the circuit. The *speed* in which the overcurrent device opens is critical for the protection of the equipment.

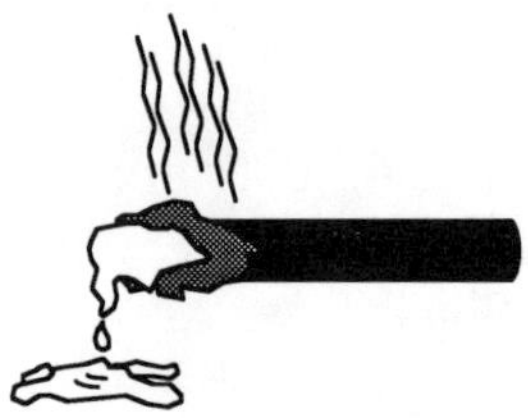

Whereas overload currents occur at rather modest levels, the short-circuit or fault current can be many hundreds of times larger than the normal operating current. A high level fault may be 50,000 amps or more. If not cut off within a few thousandths of a second, there can be severe damage, melting of conductors, arcing and fires.

The point to point method of calculating short-circuit currents is based on computation of the two main circuit impedance parameters, transformers and conductors. Of these two components, the transformer is generally the major short-circuit current factor for faults near the service entrance. The percent of impedance of the transformer can vary considerably. The transformer specification of the drawing should always be checked.

Transformer impedance (Z) helps determine what the short circuit current will be at the transformer secondary. Transformer impedance is determined as follows:

The transformer secondary is shorted. Voltage is applied to the primary which causes full load current to flow in the secondary. The applied voltage divided by the rated primary voltage is the impedance of the transformer.

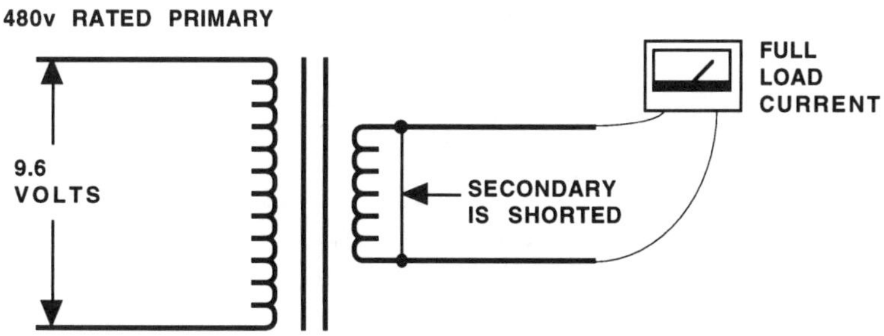

For a 480 volt rated primary, if 9.6 volts causes secondary full load current to flow through the shorted secondary, the transformer impedance is 9.6/480 = .02 = 2% impedance.

If there are transformers in the system the table below is a quick way to determine the short-circuit current available at the terminals.

$$\text{Formula for short-circuit current} = \frac{\text{F.L.A.} \times 100}{\%Z}$$

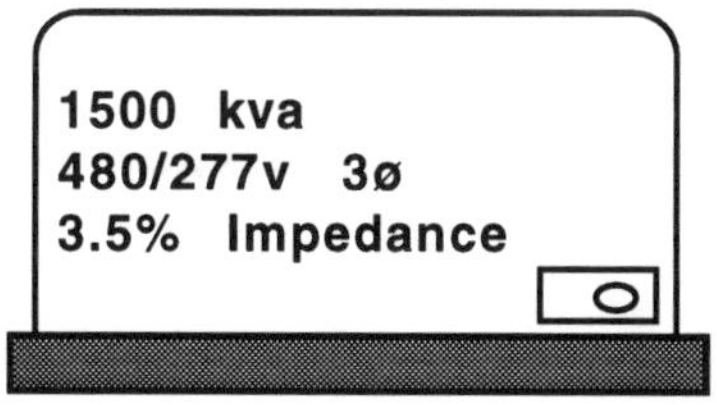

The short-circuit current is 51,543 amperes for a 1500 kva (1804 FLA) 480/277v 3ø transformer with 3.5% impedance. This number was derived by using the above formula.

$$\frac{1804\text{a} \times 100}{3.5\%} = \frac{180400}{3.5} = 51{,}543 \text{ amperes}$$

To find the worst case scenario divide by .9.

Example: 51,543/.9 = **57, 270 AMPERES WORST CASE SCENARIO**.

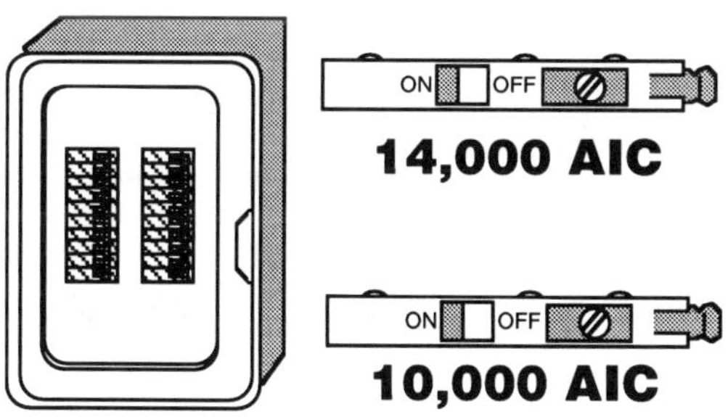

For a fully rated panelboard, the short-circuit current rating of the panelboard is NO more than the **LOWEST RATED DEVICE** installed in it. The short-circuit rating of this panelboard would be 10,000 amperes.

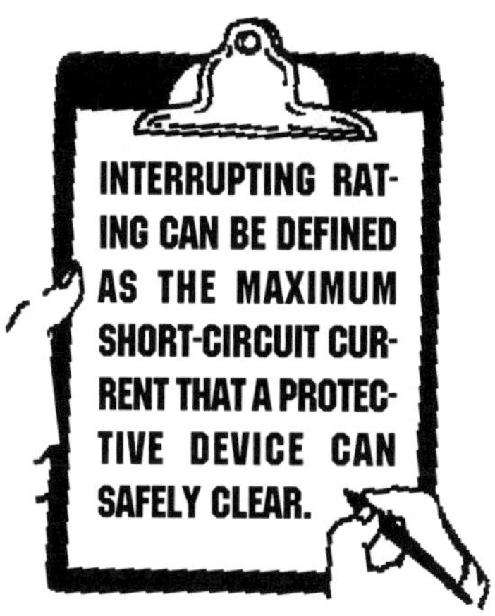

FAULT CURRENTS

To properly apply fuses or breakers within their interrupting ratings it is necessary to know the available currents on the line side of the fuses or breakers.

SINGLE PHASE 1ø			SHORT CIRCUIT
KVA	FLA	IMPEDANCE	AMPS
25	104	1.6	11,431
37.5	156	1.6	16,961
50	209	1.7	21,065
75	313	1.6	32,789
100	417	1.6	42,779
167	695	1.8	60,038

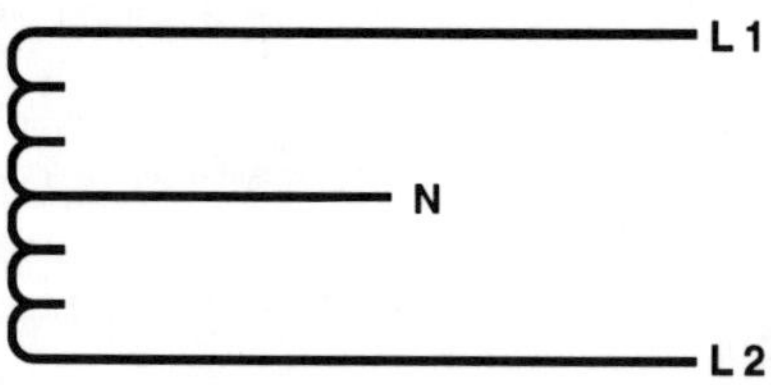

A line to neutral fault involving one of the secondary half-windings of a single-phase transformer allows approximately twice as much short-circuit current to flow as does a line to line fault involving the full secondary winding impedance. Consequently fuse or breaker selections for a 3-wire service must be based on the faulted half-winding value of short-circuit current.

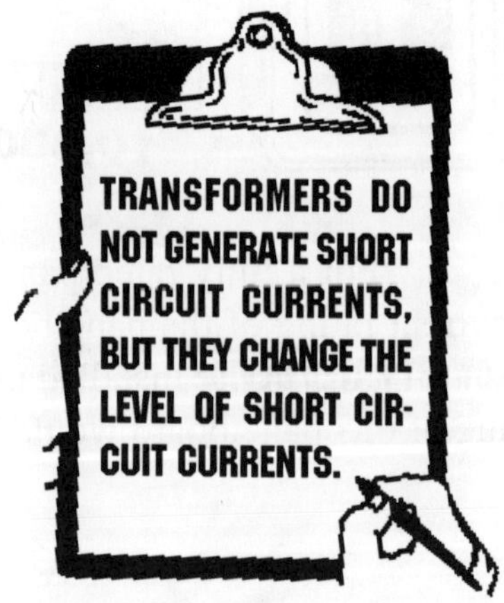

FAULT CURRENTS

THREE PHASE 3ø KVA	FLA	IMPEDANCE	SHORT CIRCUIT AMPS
150	417	2.0	20,850
225	625	2.0	31,250
300	834	2.0	41,700
500	1388	2.0	69,400
750	2080	3.5	59,429
1000	2776	3.5	79,314
1500	4164	3.5	118,971
2000	5552	5.0	111,040
2500	6950	5.0	139,000

THREE PHASE 3ø KVA	FLA	IMPEDANCE	SHORT CIRCUIT AMPS
112.5	135	1.0	13,500
150	181	1.2	15,083
225	271	1.2	22,583
300	361	1.2	30,083
500	601	1.3	46,231
750	902	3.5	25,771
1000	1203	3.5	34,371
1500	1804	3.5	51,543
2000	2406	5.0	48,120
2500	3007	5.0	60,140

Neglecting conductor impedances simplifies short-circuit calculations in circuits.

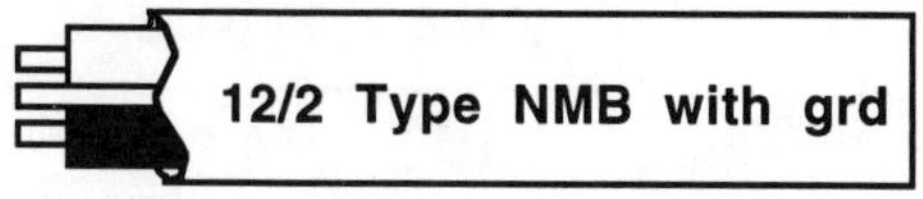

The 120 volt branch circuit shown below is 50 feet in distance from the source to the load. This circuit would contain 100 feet of conductor. A #12 solid conductor has a resistance of 1.93Ω per thousand feet. 1.93Ω x .050' = .0965Ω per 50 feet or .193Ω for 100 feet. The load has a resistance of 7.5Ω. The total resistance in the circuit is .0965 + .0965 + 7.5 = 7.693Ω.

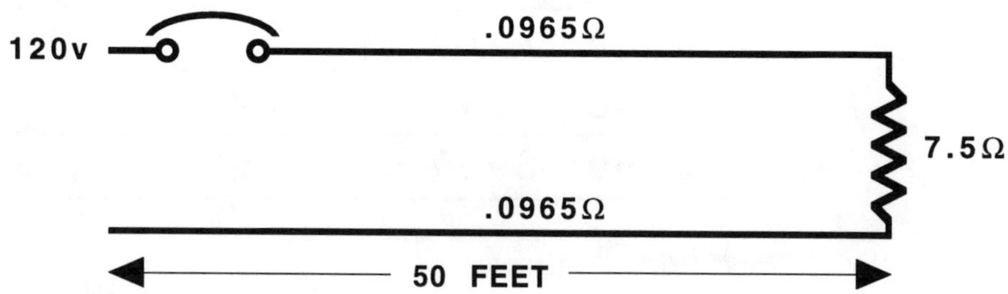

To find the current flowing in this circuit divide 120v by 7.693Ω = 15.6 amps.

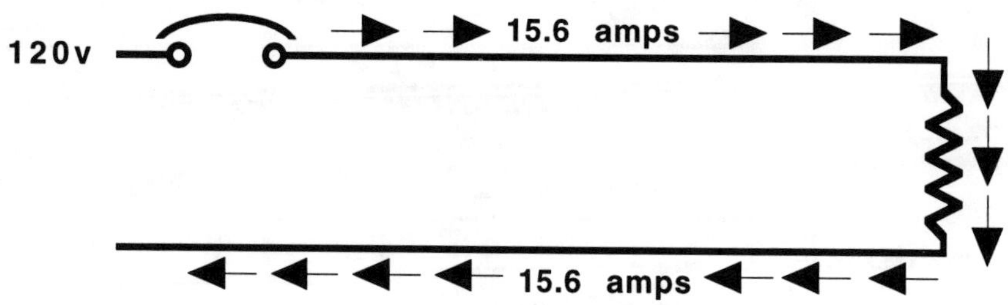

The 15.6 amps is the normal current. The 20 amp current rating of a breaker is the current level which the breaker is designed to carry continuously. The second type of rating given to a breaker is the interrupting capacity (AIC). It is the highest short-circuit current at a rated voltage that the breaker can safely interrupt. If a fault current exceeds the interrupting rating of the breaker, it can be damaged or destroyed. Severe equipment damage and personal injury can result.

Short-circuits are a result of an abnormal occurrence which causes the current to take a path outside the usual conducting path through the resistance of the load.

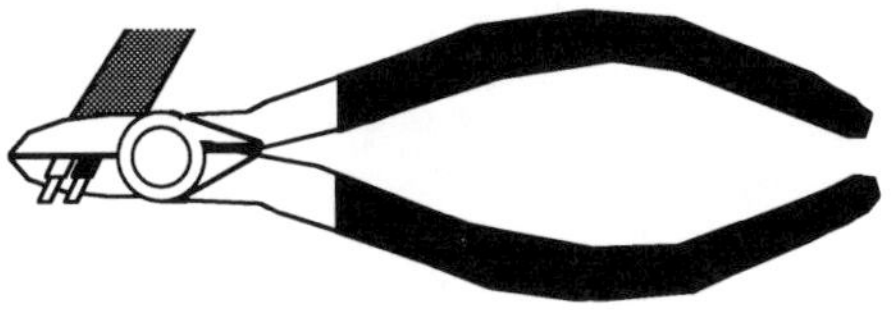

An example an electrician can relate to would be cutting into the black and white wire at the same time with the power on.

If the wire was accidentally cut at the load, the circuit would only have the resistance of the wire, as the load resistance is bypassed through the pliers which has less resistance.

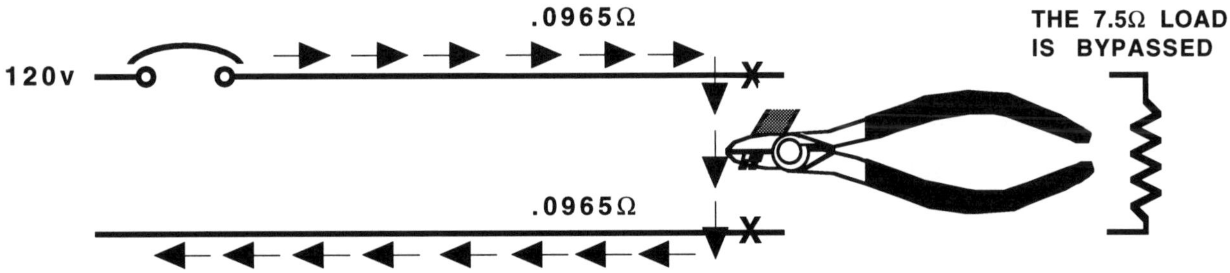

The resistance of the circuit is .0965Ω + .0965Ω = .193Ω. I = 120v/.193Ω = 622 amps.

If the wire was cut 4 feet from the breaker, the circuit would see over 7000 amperes!

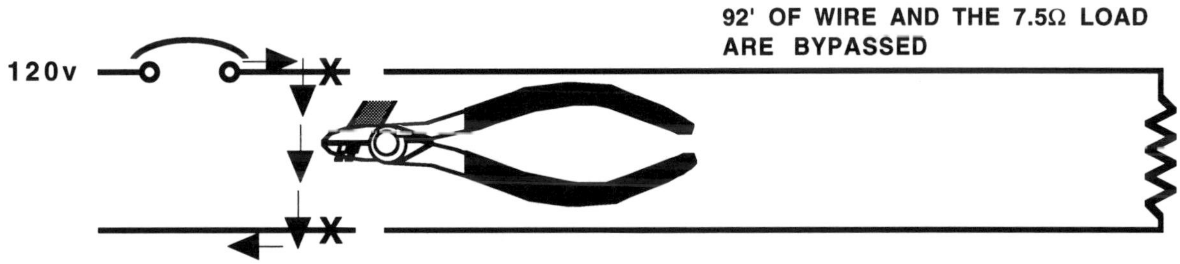

The resistance of the circuit is 1.93Ω per thousand feet x .008 feet of wire = .01544Ω.

I = 120v/.01544Ω = 7,772 amperes.

Because the load is bypassed, a very large magnitude of fault current can occur. In recent years many utility companies have increased their generating capacity and consequently, available short-circuit currents have also increased. Newer transformers with lower impedance ratings have allowed for greater magnitudes of fault current. These factors made the proper selection of an overcurrent device with a proper interrupting rating extremely critical.

A ground fault is a phase (line) conductor contacting the metal raceway enclosure (conduit), motor frame, etc.

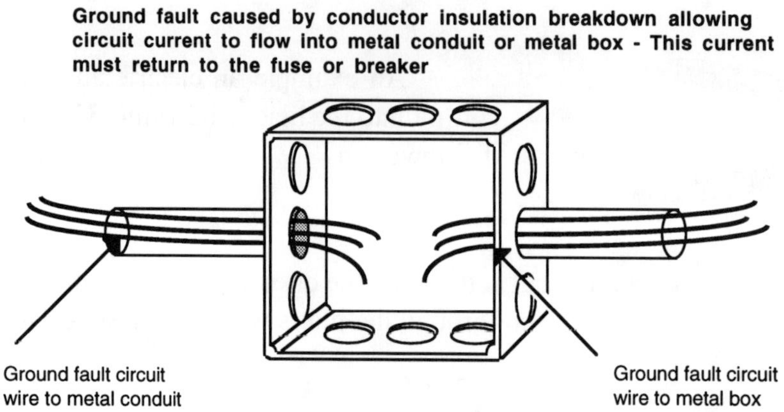

A fault may start as a ground fault and develop into a short-circuit or vice versa. The first situation is the more common. It is very important to understand the difference between a ground fault and a short-circuit as it is a more serious condition to protect against an insulation failure of a conductor to a metal enclosure.

When a fault occurs in a circuit, it is only during the time period while the fault exists that a potential hazard is present. That is why it is so important to clear the fault as quickly as possible by blowing the fuse or tripping the circuit breaker.

The fault current must return to the overcurrent protection (fuse or circuit breaker) in order to blow the fuse or trip the breaker. The grounding circuit is the return path to the overcurrent device.

Without this grounding circuit, a fault current will generate heat and eventually enough heat damage will cause the insulation to break down causing a short circuit which will cause the fuse to blow or the circuit breaker to trip.

A ground fault that is not allowed to clear itself is a very dangerous situation. A fault current may exist for minutes, hours, or days. During this time that the fault is uncleared, faults can cause a fire or even a fatal shock. Electrical accident statistics have proven that many personal injuries are caused by electric shock as a result of making contact with metal that is not normally energized. An effective grounding path would eliminate these injuries.

The interrupting rating of most branch circuit, molded case, circuit breakers like a 20 amp used in residential services are 10,000 amperes. The rating is usually expressed as "10,000 AIC" (AIC is the abbreviation for "amperes interrupting capacity.") Larger circuit breakers may have ratings of 14,000 or higher. Current-limiting fuses have an interrupting capacity of 200,000 amperes.

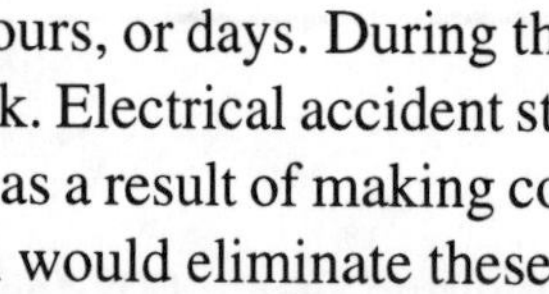

20 AMP RATED BREAKER

AIC 10,000 AMPERES

To adequately protect components, it is very important that the overcurrent protective device be current-limiting. Current-limiting overcurrent devices cut the short-circuit current off before the first full cycle can reach its maximum value.

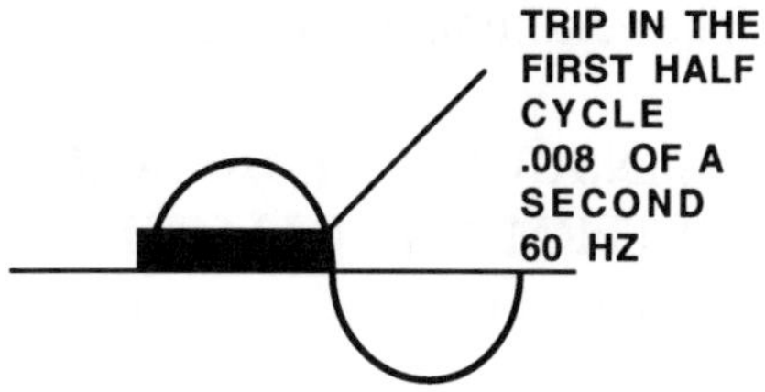

If an overcurrent device cuts off a short-circuit current in less than one-half cycle, before it reaches its total available (and highly destructive) value, the overcurrent device is a "current-limiting" device.

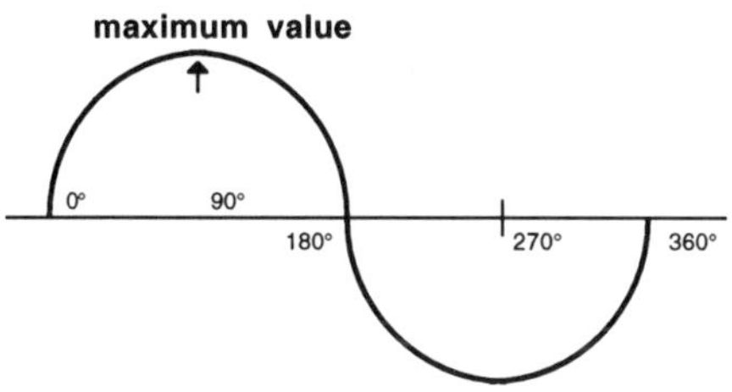

With a ground fault condition, the higher the faulted current flows the quicker the breaker will trip. If the breaker does not trip quickly and the current remains for several cycles you are generating heat which will cause damage.

The chart below shows the approximate current that conductors can carry during a ground fault or short-circuit condition without damage to the insulation.

Copper, 75°C Thermoplastic Insulated Cable Damage Table

Copper Wire Size 75° C Thermo-plastic	Maximum Short-Circuit Withstand Current In Amperes For					
	1/8 Cycle	1/4 Cycle	1/2 Cycle	1 Cycle	2 Cycles	3 Cycles
#14	4800	3400	2400	1700	1200	1000
#12	7600	5400	3800	2700	1900	1550
#10	12000	8500	6020	4300	3000	2450
#8	19200	13500	9600	6800	4800	3900
#6	30400	21500	15200	10800	7600	6200
#4	48400	34200	24200	17100	12100	9900

A #12 conductor with 75°C insulation can withstand 3800 amps, but only for 1/2 of a cycle and on a 60 cycle system that equals eight thousandths of one second, which is a very short time period.

Available short-circuit current and clearing time of the overcurrent device must be considered so the withstand rating of the conductor is not exceeded.

The 75°C conductor has three withstand classifications:

(1) INSULATION DAMAGE 150°C or 302°F
(2) TERMINAL DAMAGE 250°C or 482°F
(3) MELTING 1083°C or 1981°F

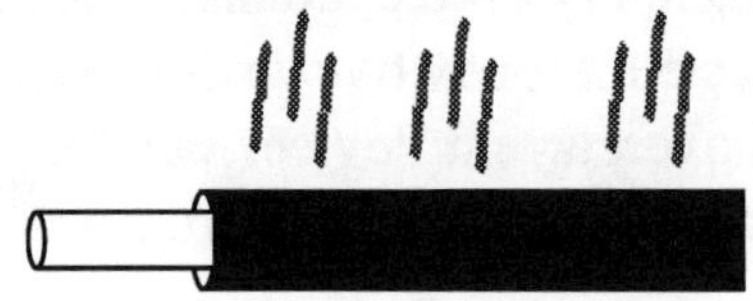

Shown below are examples using a #6 THW (75°C) conductor which has a normal ampacity of 65 amps.

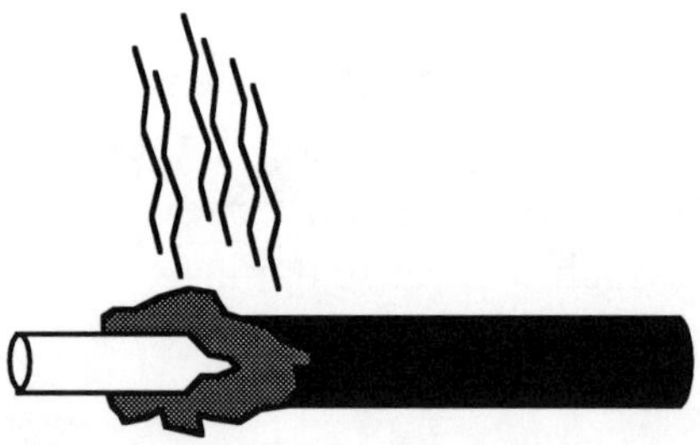

INSULATION DAMAGE

#6 THW can withstand 15,218 amps for 1/2 cycle

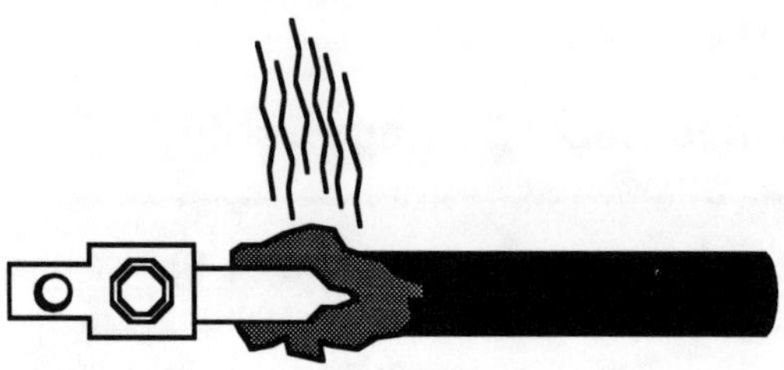

TERMINAL DAMAGE (loosening of lugs)

#6 THW can withstand 22,090 amps for 1/2 cycle

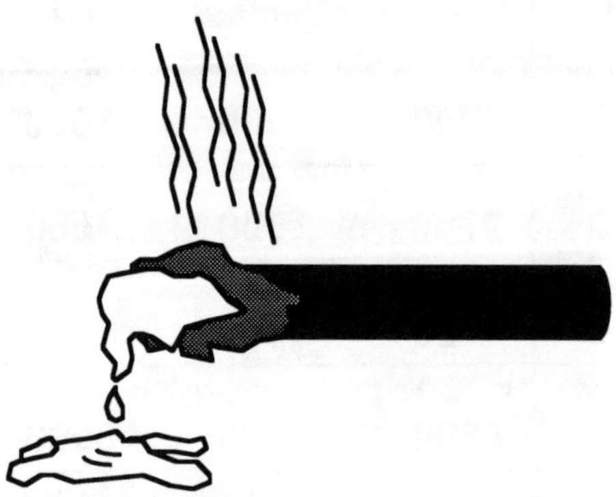

MELTING

#6 THW can withstand 39,704 amps for 1/2 cycle

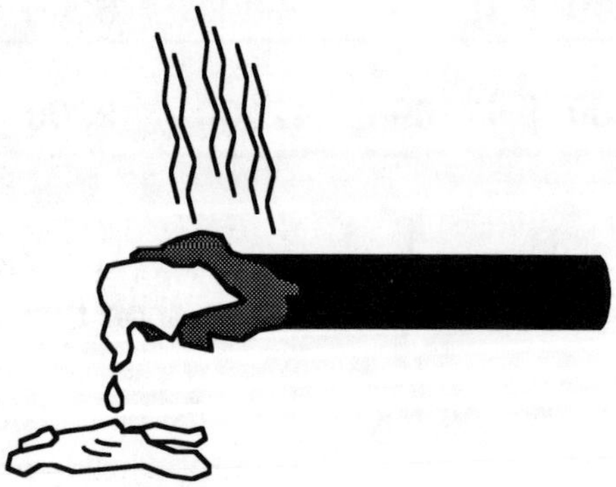

IF THE CURRENT CONTINUED FOR 5 SECONDS (300 cycles) THE #6 THW WOULD MELT AT 1620 AMPS

Insulation breakdowns to ground can occur at any point in the system where phase conductors are in close proximity to a grounded reference. The contact between the phase conductor and ground is usually not a good metallic contact, but rather includes an arcing path in air or across an insulating surface, or a combination of both.

Based on the normal rate of deterioration, insulation can be expected to have a useful life of 20 to 30 years. A fault, only for a few seconds, could cause severe life reduction.

Conductors require protection against physical damage as well as from overloads and short circuit thermal damage. Conductors are required to be protected from mechanical hazards, adverse ambient temperatures, and attack by foreign elements. Isolation is the most effective form of mechanical protection.

An example of an adverse ambient temperature is underground installations where repeated changes of high and low temperatures can fill conduits with water produced by breathing and condensation. Suitable drains at low points will remove the water as it collects.

Should heating be abnormal, as evidenced by discoloration of terminals, deterioration of molded material, or if there is nuisance tripping and the current is less than the overcurrent device rating, all bolted connections and contacts should be examined and tightened in accordance with the manufacturer's recommendations.

It is essential that each joint is bolted properly to the manufacturer's recommended torque values. **THE IMPORTANCE OF THIS STEP CANNOT BE OVEREMPHASIZED**. Periodic inspections to ensure that the joint integrity has not been affected by creep, vibration, heat cycling, or accumulation of dust or foreign matter should be performed as part of normal preventive maintenance program.

Circuit breakers after a period of time should be electrically tripped to assure proper operation. Experience has indicated that if allowed to remain in service for an extended period of time without an electrical operation, the internal mechanism and joints may become stiff so that the circuit breaker operates improperly when subjected to normal current. It is suggested that circuit breakers be subjected to a test current equivalent to 300% of the breaker trip rating and that the tripping time be measured.

CHAPTER 6
EXAM

FAULT CURRENTS EXAM

1. A #10 THW conductor can withstand ____ amperes for three cycles.

(a) 3900 (b) 2450 (c) 3000 (d) 6020

2. A current-limiting overcurrent device is required to open the short-circuit current in less than ____.

(a) one-half cycle (b) one cycle (c) two cycles (d) three cycles

3. Calculate the maximum fault current available at the secondary terminals of a 750 kva 3ø transformer, 240 volts, with 5 1/2% impedance.

(a) 24,600 amps (b) 32,800 amps (c) 180,400 amps (d) 1804 amps

4. For a 240 volt rated primary, if 8.4 volts causes the secondary full load current to flow through the shorted secondary, the transformer impedance is ____ %.

(a) 5 (b) 3.5 (c) 1.6 (d) 1.2

5. Most modern current-limiting fuses have an interrupting rating of ____ amperes.

(a) 50,000 (b) 100,000 (c) 150,000 (d) 200,000

6. The interrupting rating of a typical circuit breaker used in a residence is ____ amperes.

(a) 10,000 (b) 12,000 (c) 14,000 (d) 22,000

CHAPTER 7

POOLS

SWIMMING POOLS

There have been some terrible accidents from electrical shocks around swimming pools and fountains.

This is a very important safety inspection and the rules must be followed. The biggest asset the inspector has is all equipment must be *listed*. This must be enforced.

It is important to understand the definitions of pool terms as the Code rules may differ.

A permanently installed pool is capable of holding water deeper than 42 inches.

A pool is for swimming or wading. A fountain is for cosmetic or aesthetic reasons.

A permanently installed pool can be one installed above ground as long as it is capable of holding water deeper than 42 inches.

A storable swimming pool is one that has a maximum depth of 42 inches.

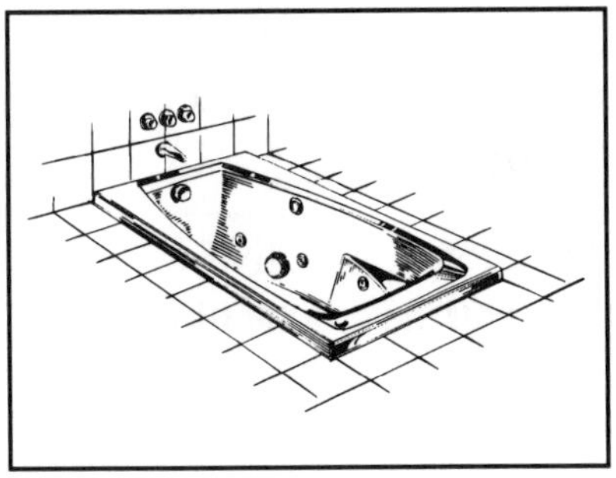

A hydromassage bathtub is a permanently installed bathtub equipped with a recirculating piping system. The water is usually drained out after use.

Most plumbing codes require a removeable access panel for servicing or replacing the pump motor.

A spa or hot tub are hydromassage units designed for people to get into, and generally the water *won't be* drained so they contain a filtering system along with heaters and blowers.

Being in a concrete pool of water in contact with earth, with an electrical light installed in the water could present a severe electrical shock to a person if the pool electrical equipment does not meet the minimum standards of the Code.

Some people don't realize that even though radios, hair dryers, hair curlers, etc. are turned "off", if they fell into bath water they could prove lethal.

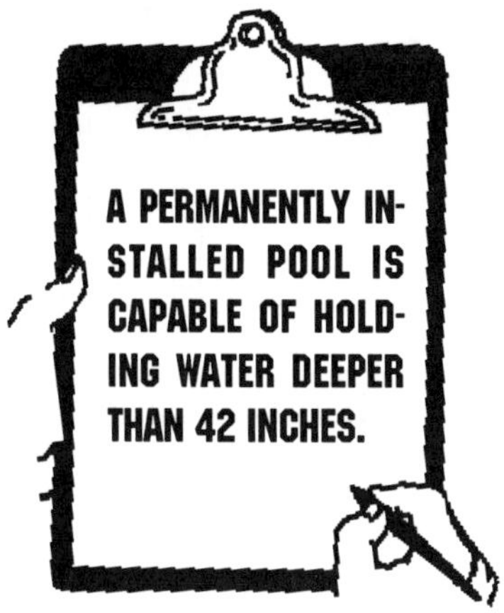

With respect to water a person can receive an electrical shock from a difference of potential (voltage) in the water with respect to ground (earth), or the existence of a potential gradient in the water itself, or a difference in potential between various points on a surface.

The shock experienced from a person in a pool is different from touching a live wire. The pool water makes contact with the entire skin surface of the body rather than the touch point of your finger to a live wire.

From working as an electrician one has most likely experienced shock from touching a "hot" wire or a metal enclosure that is energized. The picture below shows *voltage gradients* as a radio plugged in a 120v receptacle falls into the swimming pool.

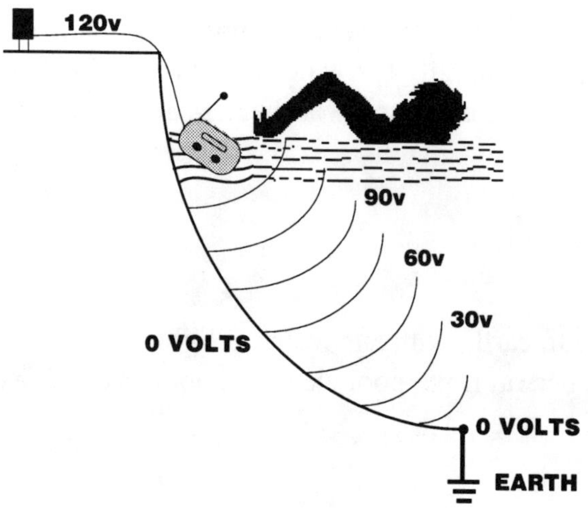

These rings of voltage radiate out from the pool walls from 120v at the radio to 0 volts at the grounded wall.

Gradient voltages are found in the entire swimming pool water in varying voltages.

A swimmer subjected to voltage gradients as low as 4 volts in pool water can be immobilized by the disabling effects of current passing through the brain from water in the ears, nose and other body openings.

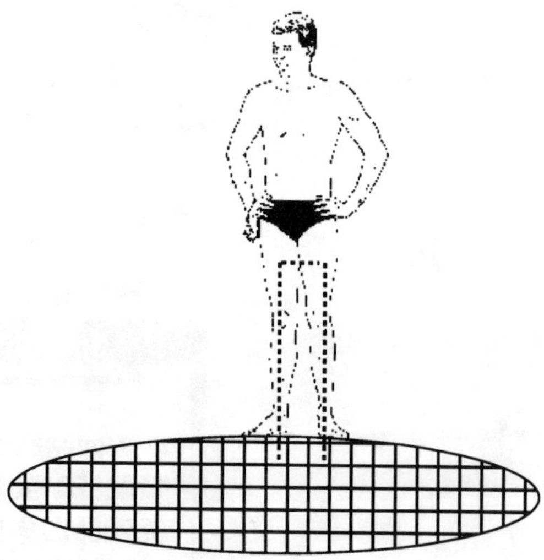

If a difference of potential (voltage) occurs between two points on the pool slab, a current can flow from one foot through the legs to the other foot completing the circuit.

BONDING and GROUNDING

Bonding and grounding are different and you must first understand the difference. Bonding is the connecting together of all *metal* surfaces to limit the possibility that the surfaces might be energized at different voltages, thus creating a shock hazard for a person touching two metal surfaces at the same time. Bonding minimizes voltage differences between exposed *metal* parts.

Grounding is extending the bonding circuit to the grounding electrode by means of an *equipment grounding* conductor to bring everything within touch to the same potential; *earth* potential. Grounding provides a low-resistance path to ground for the fault currents. If a metal object would become energized because of an insulation failure proper grounding will open the overcurrent device.

All metal parts that are within 5' of the inside walls of a pool must be bonded together with a reliable corrosion-resistant electrical connection unless separated by a permanent barrier. This is to eliminate shock hazard from any stray currents that may be induced to the metal parts around the pool.

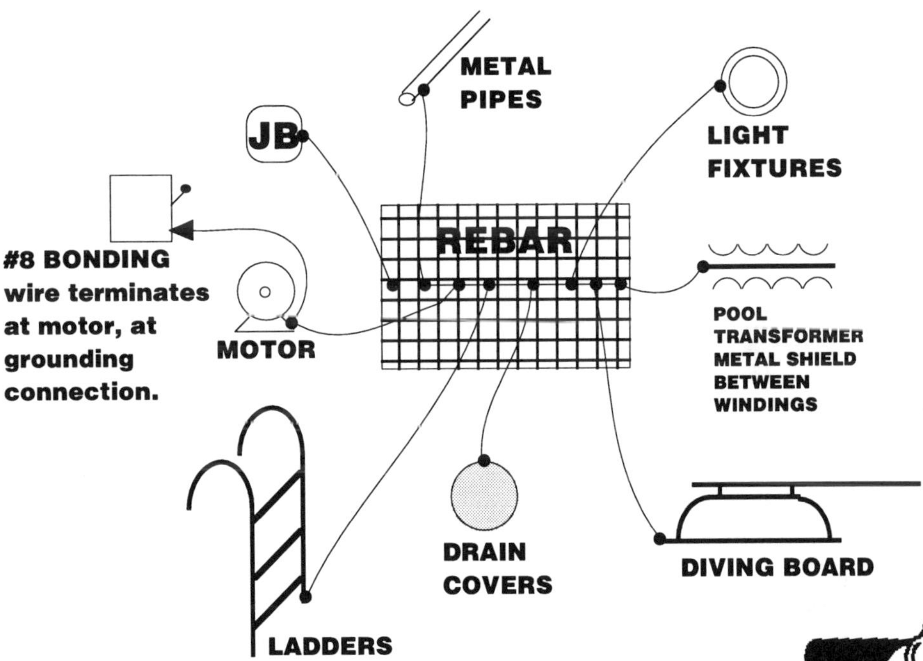

The bonding conductor is required to be a #8 solid minimum. It may be bare, covered or insulated, and may be directly buried. Bonding conductors if insulated do not follow any color coding rules. The equipment grounding conductor is green. The bonding conductor has no color indentification rules. Where connections are made below grade or embedded in concrete, the connectors (brass, copper, or copper-alloy) must be marked to show suitability for the application.

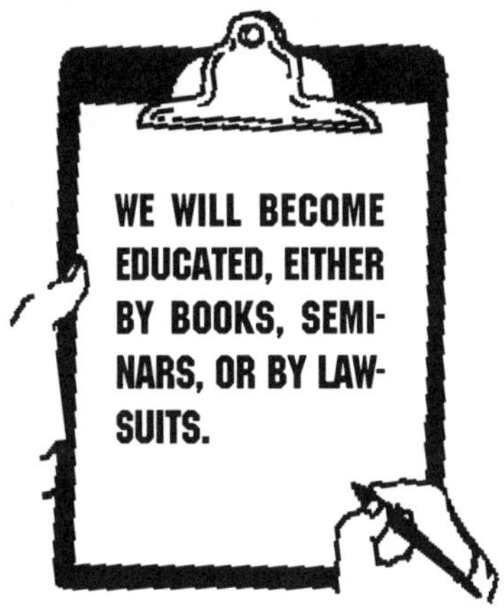

The sketch below would *not* be recommended as the way to connect the metal parts. The loosening at one of the parts would open the bonding path.

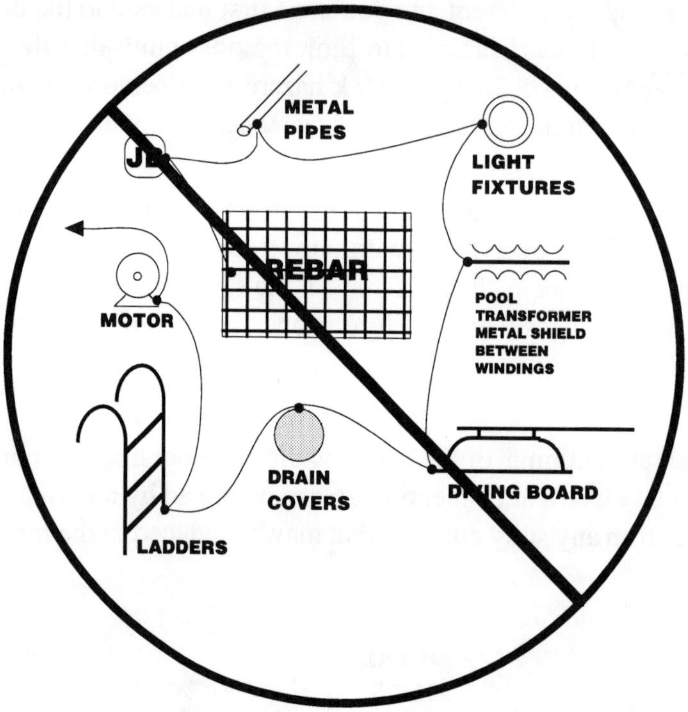

All connections are to be inspected *prior* to the pouring of concrete. The #8 copper in some cases may have to be connected to a ladder made of aluminum or stainless steel. The connector must be *listed* suitable for this use.

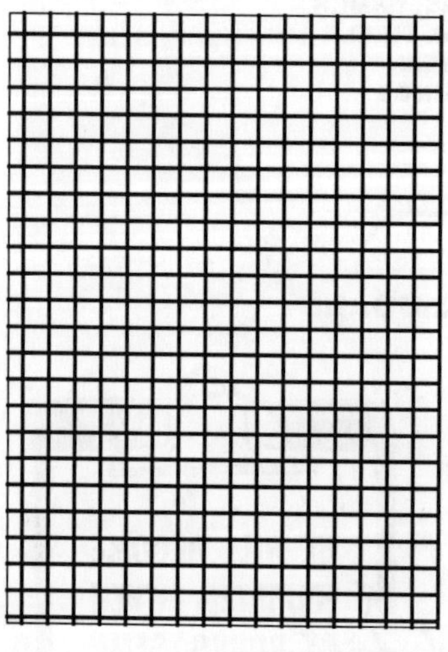

The rebar tie wires are considered a suitable connection to connect the rebars together provided they are tight.

Generally, the center-line rebars are bonded together and bonded to the #8 bonding conductor at several points.

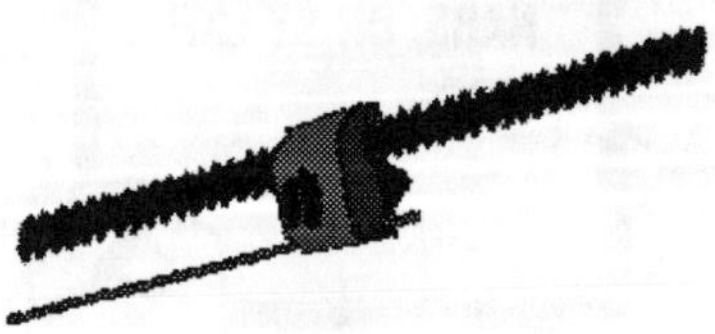

The #8 solid copper conductor is connected to the common bonding grid using brass or copper connectors or clamps to prevent corrosion of this very important connection.

GROUNDING

The objective of grounding is to provide a path to the grounding electrode at the service and to assure a low-impedance path for fault currents to flow back to the grounded neutral to permit proper operation of the overcurrent devices. Grounding puts all metallic parts to earth potential of zero reducing shock hazard.

Grounding requires running an insulated *green* equipment grounding conductor with the circuit conductors to junction boxes, lights, motors, transformer enclosures, etc. In general, metal raceways are not used to provide a grounding path for swimming pool systems. And rightfully so, a metal conduit system can open due to corrosion. This has been a problem in pump houses where the conduits are exposed to chlorine or acids. This is the same reason the equipment grounding conductor is required to be *insulated* and not bare.

Bonding conductors do not have to be in conduit. Grounding conductors do. Bonding conductors connect directly to enclosures whereas grounding conductors must terminate at an equipment grounding terminal.

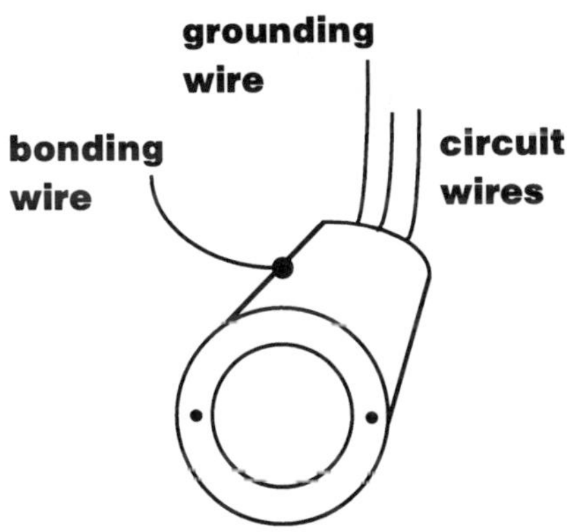

Grounding conductors run with the circuit conductors.

The grounding conductor to a wet-niche, dry-niche, or no-niche lighting fixture shall not be smaller than a #12. It shall be an *insulated* copper conductor, shall be installed in rigid metal, IMC, or rigid PVC (Schedule 40 or 80). EMT and ENT are permitted to be used where installed on or within a building.

The metal parts shown below don't have any circuit wires, all they have is a bonding wire which connects directly to the metal part as there is no terminal to terminate at.

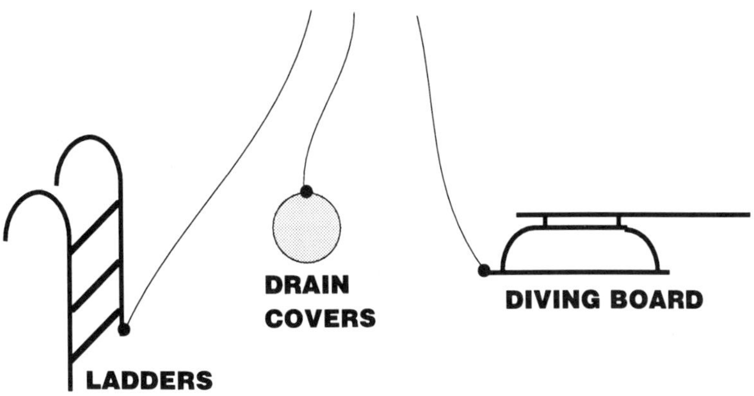

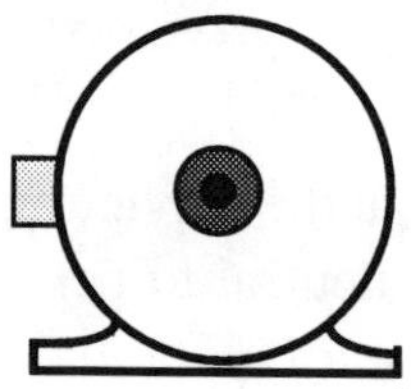

Pool-associated motors shall be connected to an insulated copper equipment grounding not smaller than #12. The grounding conductor along with the circuit conductors shall be installed in rigid metal conduit, IMC, rigid PVC (Schedule 40 or 80), or Type Metal Clad cable *listed* for the application. UF or USE cable is *not* permitted.

Exceptions permit the use of EMT where installed on or within a building. Where flexibility is needed at the motor, liquidtight flex or nonmetallic conduit with approved fittings is permitted.

NM cable (romex) is permitted for the *part* of the motor circuit that is run in the *interior* of a one family dwelling. BX cable is *not* permitted as the minimum equipment grounding conductor size is #12.

Another exception permits the use of flexible cord for *cord and plug connected equipment.* For other than storable pools, the cord shall not exceed *3' in length* and contain an equipment grounding conductor size #12 minimum. The cord must be *listed* for outdoor use. Plug and cord connected motors facilitate the removal or disconnection of the motor for maintenance or repair.

A receptacle installed in a wet location shall be in a weatherproof enclosure, the integrity is *not affected* when the attachment plug is inserted. This is an area I see often violated. The cover for the receptacle must remain weatherproof with the cord plugged into the receptacle.

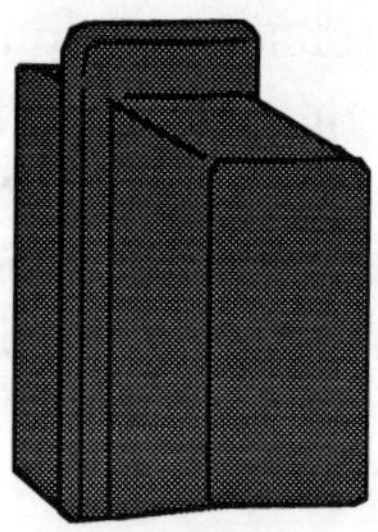

HOODED COVER IS REQUIRED

Pool cover motors shall be located at least 5' from the inside wall of the pool unless separated from the pool by a wall, cover, or other permanent barrier. Motors installed below grade level shall be of the totally enclosed type. The motor and controller shall be connected to a GFCI circuit.

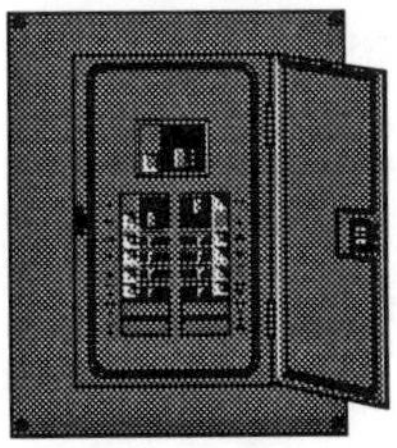

An insulated equipment grounding conductor sized from Table 250-122 and not smaller than #12 is required between the service equipment and the panelboard that supplies circuits to the pool equipment.

The required insulated equipment grounding conductor along with the circuit conductors are to be installed in rigid metal conduit, IMC, or rigid PVC (Schedule 40 or 80). EMT is permitted where run on or within the building.

For an **existing** panelboard, an exception permits flexible metal conduit or an approved cable assembly having an insulated *or covered* equipment grounding conductor not smaller than #12.

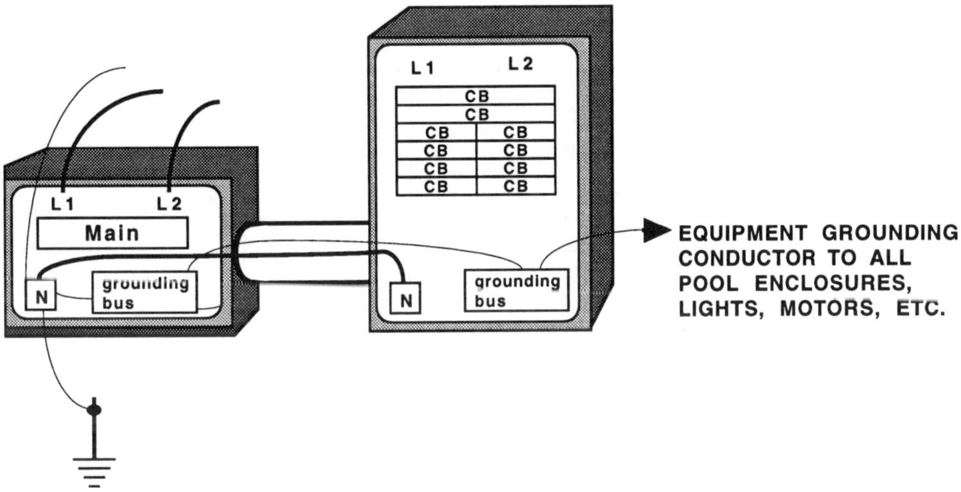

The equipment grounding conductor is to be run without joint or splices means it is to be connected to an equipment grounding *terminal* from the panelboard to the equipment grounding *terminal* at the equipment. Wire nuts are not permitted for connecting the equipment grounding conductor.

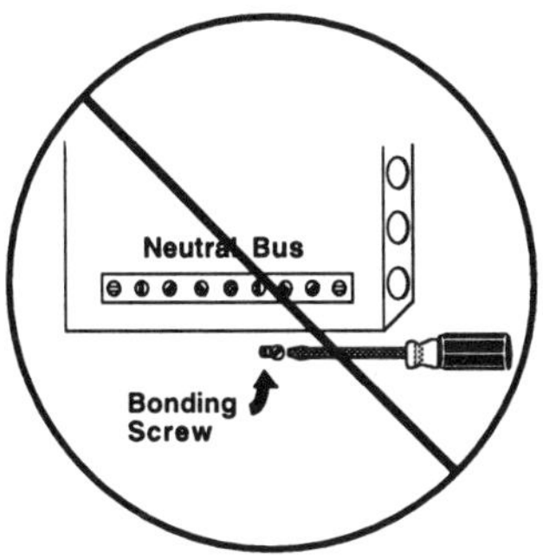

Do **not** tie the equipment grounding and neutral together at a subpanel, they are connected only at the **service** equipment, not a remote panel. Such a connection ties the grounding system in parallel with the neutral and can cause GFCI protective devices not to work. Also, any voltage above ground in the neutral due to unbalanced currents will appear on the equipment ground, and these voltages can be very dangerous in a pool area.

S W I M M I N G P O O L S

Receptacles shall be at least 10' from the inside walls of a pool or fountain and *shall be GFCI protected.*

At a dwelling a permanently installed pool requires at least one receptacle located between 10' and 20' from the inside wall.

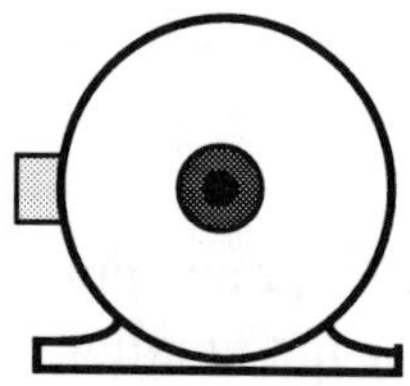

Most pump motors are hard wired, but in some cases in freezing conditions the motors are plug and cord connected for easy removal in the winter.

An exception permits a receptacle for a pump motor to be between 5' and 10' from the inside walls. The receptacle shall be a *single* receptacle of the *locking* and grounding type and *shall be GFCI protected.*

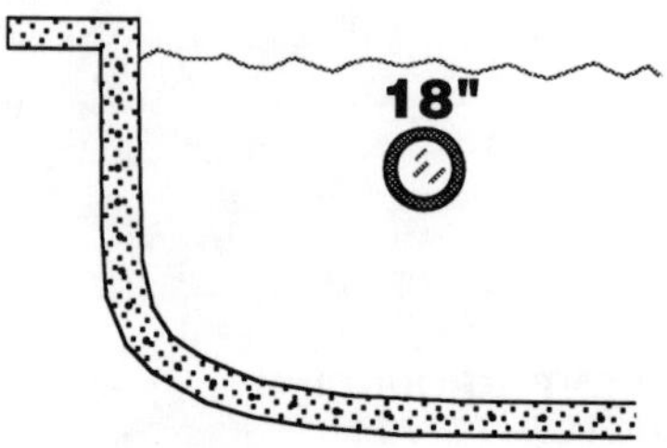

The top edge of the underwater fixture is to be at least 18" below the normal water level.

Underwater lighting fixtures at more than 15 volts must be approved and *GFCI protected.*

The aging process, corrosion, and questionable original wiring have made underwater swimming pool lighting fixtures a potential source for electrocution.

The Code requires the inside forming shell termination of the #8 insulated grounding wire along with the other wires to be encapsulated with a listed potting compound to protect from deterioration if pool water should enter.

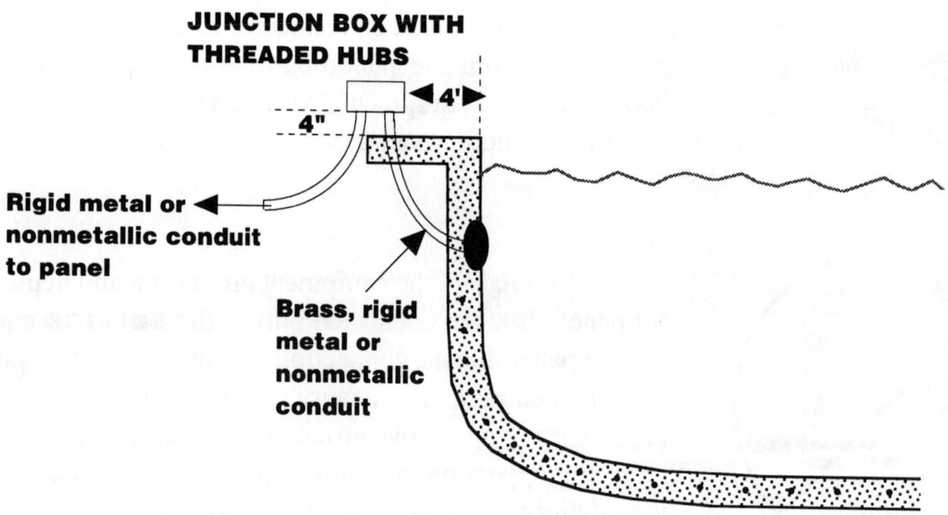

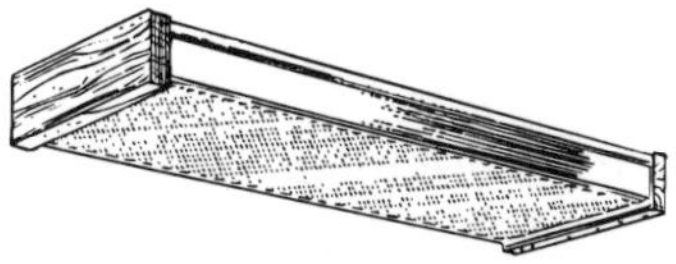

The general rule is lighting fixtures or lighting outlets, or ceiling fans shall not be installed above a pool, or over the area extending 5', measured horizontally from the inside walls of the pool, unless the fixtures are mounted at least 12' above the maximum pool water level.

There are *exceptions* for existing fixtures, totally enclosed fixtures GFCI protected, etc.

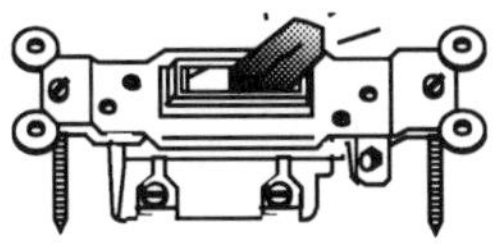

Wall switches must be located at least 5' from the inside walls to eliminate possible shock hazard to persons in the water.

Overhead wiring must be never less than 10' over the pool, diving board, platforms, towers, etc.

Telephone and cable TV lines are also permitted above the pool but not less than 10' including the support messengers.

Underground wiring other than *pool associated* shall not be within 5' horizontally from the inside wall of the pool. An exception permits underground wiring closer than 5' where installed in a rigid corrosion resistant raceway and suitable for the location.

Pool accidents have occured from the use of long vacuum skimmer poles, with lightweight aluminum handles, used in cleaning debris from the water surface and touching overhead wires.

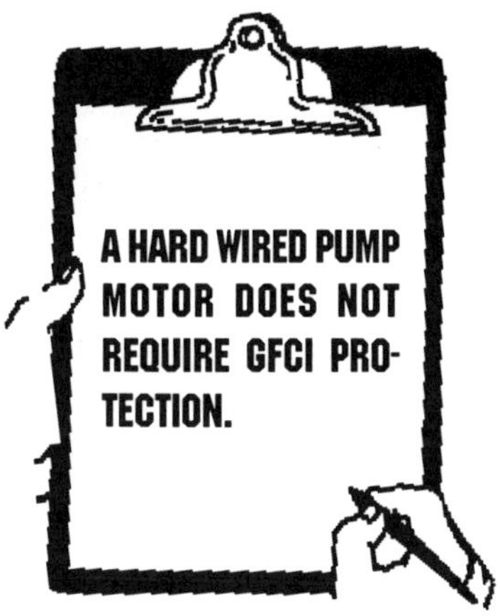

SPAS and HOT TUBS

The difference between a spa and a hot tub is generally a hot tub is made of wood and a spa is made of fiberglass, plastic or some other manmade product.

Both contain recirculating equipment and heaters. They are not drained after each use as with a bathtub.

Where located outdoors they follow most of the rules for swimming pools. An outdoor spa or hot tub has greater hazards due to corrosion and exposure to grounded surfaces.

Some spas and hot tubs come equipped with a plug and cord connection. Some are listed to be converted to hard wired. The manufacturer will furnish instructions on how to convert to hard wired. Listed units are permitted to have a cord up to 15' in length. These units must be GFCI protected.

•Note: A receptacle installed in a wet location shall be in a weatherproof enclosure, the integrity is *not affected* when the attachment plug is inserted. This is an area I see often violated. The cover for the receptacle must remain weatherproof with the cord plugged into the receptacle.

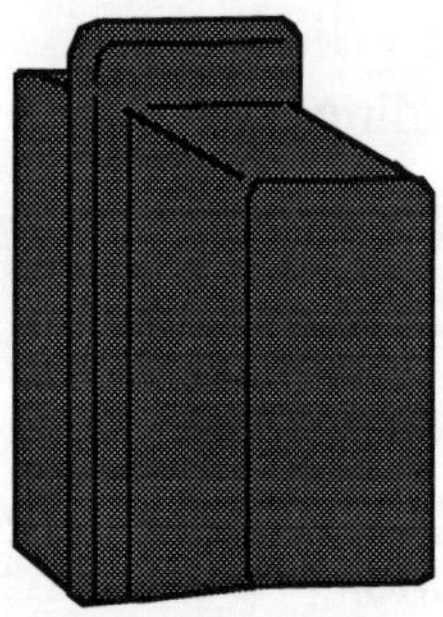

HOODED COVER IS REQUIRED

Listed units that have a factory installed remote control panel can be connected with liquidtight flexible conduit up to 3' in length.

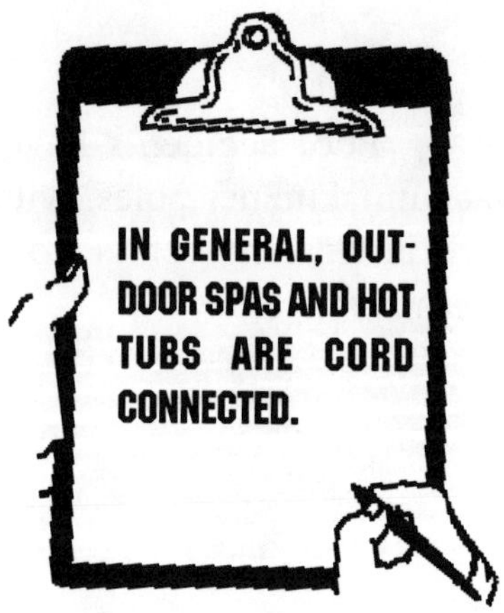

At least one receptacle outlet must be installed at least 5', but not more than 10' from the inside edge of the spa or hot tub.

Without this receptacle it was found that long extension cords were being used for appliances used in the spa and hot tub areas.

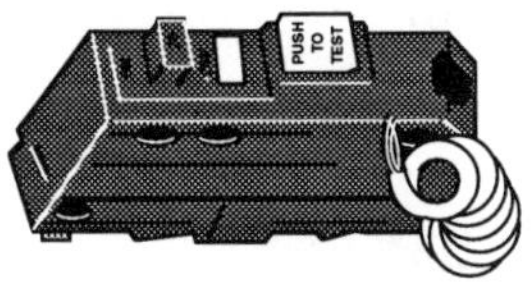

No receptacles are permitted within 5' of the spa or hot tub. All receptacles within 10' of the inside walls must be GFCI protected. Any receptacle that provides power **must be GFCI protected** regardless of configuration, amperage, voltage rating, or location.

An exception for *listed* self-contained unit or listed packaged equipment assembly that has **integral** GFCI protection would not require additional GFCI protection.

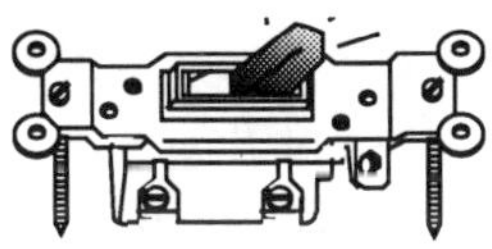

Wall switches must be located at least 5' from the inside walls to eliminate possible shock hazard to persons in the water.

Light fixtures, ceiling fans, lighting outlets located over a spa or hot tub or within 5' from the inside walls shall be a minimum of 7' 6" above the maximum water level and shall be GFCI protected.

Exceptions are: (1) If over 12' above water level shall not require GFCI protection.

(2) Recessed or surface mounted fixtures with glass or plastic lens or globes and a nonmetallic body isolated from contact and suitable for damp locations and GFCI protected can be installed less than 7' 6" over the spa or hot tub.

Code section 410-4d prohibits ceiling fans over a bathtub 3' horizontally and 8' vertically. But section 90-3 states Chapter 1 through 4 apply **except** as amended by Chapters 5, 6, and 7 for the particular conditions.

Spas and hot tubs installed indoors have different rules than ones installed outdoors. Generally outdoor installations are cord connected. Indoor installations can be cord connected if rated 20 amperes or less. Over 20 amps indoors must be hard wired using a Chapter 3 wiring method.

Bonding and grounding of spas and hot tubs are very similar to swimming pools.

Bonding: All metal fittings, equipment motors, metal pipes, conduits, metal surfaces within or attached to the spa or hot tub or within 5'. Bonding by means of threaded metal pipe and fittings, metal to metal of a frame, or by means of a #8 or larger, solid, bare, insulated, or covered bonding jumper.

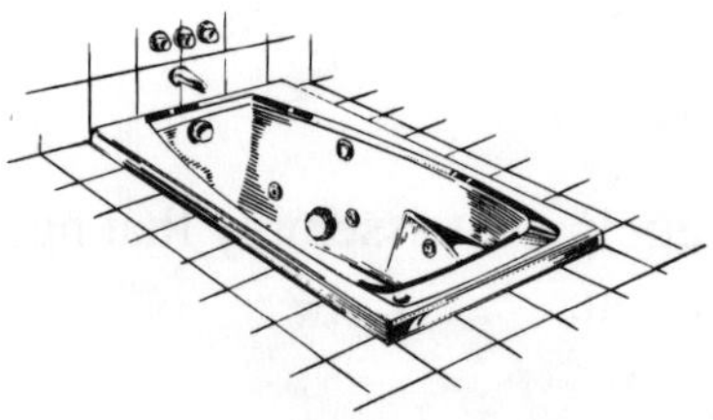

HYDROMASSAGE BATHTUBS

A hydromassage bathtub (sometimes called whirlpool tub) along with its components shall be GFCI protected. All 125v receptacles within 5' of the inside walls shall be GFCI protected.

The Code does not consider it anymore of a shock hazard than a bathtub so it can follow the same Code wiring rules.

Pools and tubs for therapeutic use basically follow the same general rules as a swimming pool.

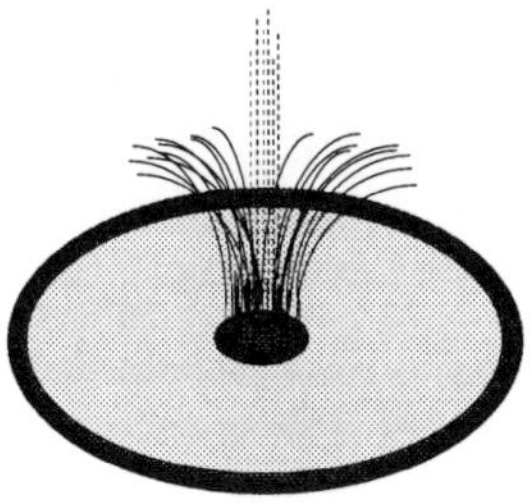

FOUNTAINS

The difference between a pool and a fountain is you swim in a pool and a fountain is a work of art and taste and with the creation and appreciation of beauty.

Fountains have special rules unless the water in a fountain is common with a swimming pool then they follow the same rules of the pool.

All fountain equipment shall be *GFCI protected*. There is an exception for equipment 15 volts or less supplied by an approved isolating transformer.

Lighting fixtures listed for fountains use the term *submersible*.

Lighting fixture lenses must be installed below the normal fountain water level unless *approved* for above water.

Fountain equipment must be installed so it can easily be maintained (brought to the surface for relamping, etc.) without lowering the level of the water. The cords shall have strain relief so the pulling of the cord won't pull on the terminals when a fixture is lifted out of the water.

Cords must be rated for hard service and marked *water resistant*. The ends of the cord, and terminations must all be in a suitable potting compound so if water would get into the cord it would not get into the equipment.

Equipment must be securely fastened in the fountain so it won't tip over.

The grounding of a fountain follows the rules of 680-25.

CHAPTER 7
EXAM

POOLS - EXAM 1

1. All 125v, single-phase receptacles within ____ of the inside walls of a hydromassage tub shall be GFCI protected.

(a) 5' (b) 6' (c) 8' (d) 10'

2. Which of the following pool parts are required to be bonded together?

I. all fixed metal parts within 5 feet of the inside walls
II. all forming shells and mounting brackets of a no-niche fixture
III. all metal parts of an underwater sound system

(a) I only (b) II only (c) I and II only (d) I, II and III

3. Wet-niche lighting fixtures shall be connected to an equipment grounding conductor not smaller than ____.

(a) #10 (b) #6 (c) #8 (d) #12

4. A pool recirculating pump motor receptacle shall be permitted not less than ____ feet from the inside walls of the pool.

(a) 5 (b) 8 (c) 10 (d) 15

5. The following pool equipment shall be grounded ____.

I. ground fault circuit-interrupters
II. transformer enclosures
III. electric equipment located within 5 feet of the inside wall of the pool

(a) III only (b) II and III only (c) II only (d) I, II and III

6. A forming shell shall be provided with a number of grounding terminals that shall be ____ the number of conduit entries.

(a) one more (b) two more (c) same as (d) none of these

7. Junction boxes for pool lighting shall not be located less than ____ feet from the inside wall of a pool unless separated by a fence or wall.

(a) 3 (b) 4 (c) 6 (d) 8

8. The maximum length of exposed cord in a fountain shall be ____ feet.

(a) 3 (b) 4 (c) 6 (d) 10

9. Overhead service drop conductors shall have a horizontal clearance of ____ feet from a pool.

(a) 8 (b) 10 (c) 15 (d) 20

10. Lighting fixtures mounted on walls shall be installed with the top of the fixture lens at least ____ below the normal water level of the pool.

(a) 15" (b) 3' (c) 18" (c) 12"

11. Metal conduit and metal piping within ____ feet of the inside walls of the pool and that are not separated from the pool by a permanent barrier are required to be bonded.

(a) 4 (b) 5 (c) 8 (d) 10

12. A pool panelboard, not part of the service equipment, shall have a grounding conductor installed between ____.

(a) its grounding terminal and a separate ground
(b) its grounding terminal and a ground rod
(c) its grounding terminal and the grounding terminal of the service equipment
(d) its grounding terminal and bonding grid

13. Ground fault circuit-interrupters shall be installed in the branch circuit supplying underwater pool lighting fixtures operating at more than ____ volts.

(a) 12 (b) 15 (c) 24 (d) 50

14. Disconnecting means shall be accessible, located within sight from pool, and shall be located at least ____ horizontally from the inside walls of the pool.

(a) 18" (b) 2' (c) 4' (d) 5'

15. What size conductor shall be connected between the ground grid and all metal parts of swimming pools?

(a) #8 (b) #10 (c) #6 (d) #4

CHAPTER 8

EMERGENCY SYSTEMS

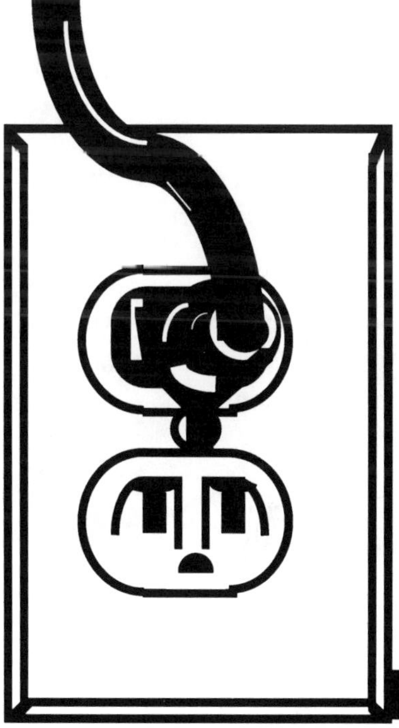

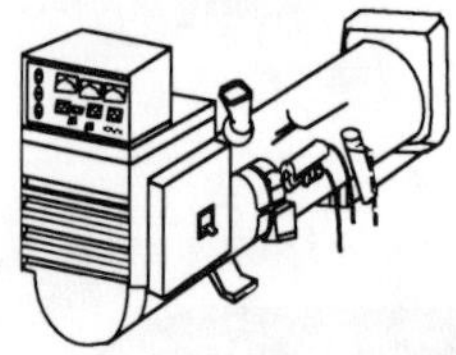

EMERGENCY SYSTEMS

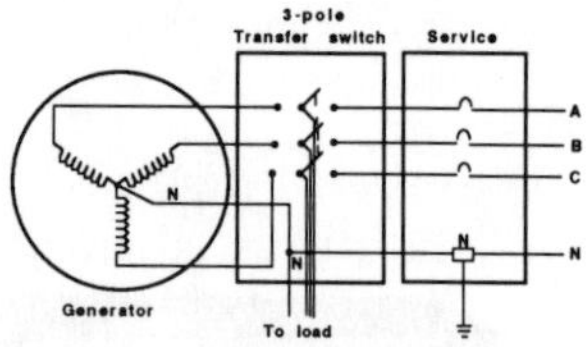

Where the failure of normal power to a building would result in a hazard to life or property, provision should be made for an emergency power supply.

Emergency systems are classified and made law by federal, state, municipal, or other code or by a governmental authority.

When such systems are installed in buildings under the electrical inspector's jurisdiction, the inspector should conduct or witness a test on the complete system upon installation and periodically afterward. The periodic tests are performed to assure that their maintenance keeps the emergency systems in proper operating condition.

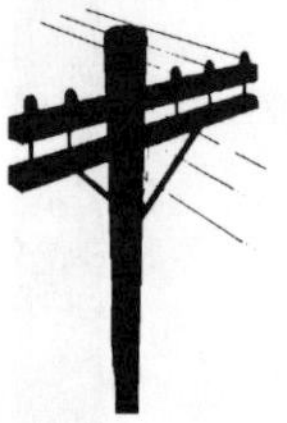

Such elements as lightning, vehicles striking power poles, ice storms, etc. are a few causes of interruptions of power from the utility company. Because utility companies have no control over these disturbances on their systems they are unwilling to accept responsibility for them or make power availability guarantees.

Short circuits, open circuits, ground faults, undersized conductors, equipment failures, operator errors, overloads, etc. are causes of interruption on the distribution side.

Emergency power and lighting is of extreme importance in hospitals and similar institutions. The required 10 second transfer time can seem like a long time during open-heart surgery or if you are on a lung machine.

When the situation is NOT life threatening standby power may be required for loss of production or damage to goods such as in a greenhouse, egg farm, blood banks, ice cream stores, etc. Windowless buildings can become an incubator without ventilation equipment.

Even in the coldest regions, a building can go without heat for a minimum of 30 minutes.

Emergency warning lights may be required on taller buildings to warn aircraft. Warning lights on edges of waterways to warn ships. These lights are to prevent injury and property damage.

The National Electrical Code itself does *not* require emergency light, exit signs, or power.

The location of exit lights is covered by the **LIFE SAFETY CODE**. The electrical code indicates how the installation will be made, but not where the emergency lighting is required, except for essential electrical systems in a health care facility.

Means of egress shall be illuminated at all times that the building is occupied, with light of not less than *one footcandle* intensity at the floor level, except theaters which shall have not less than 1/5 footcandle in aisles. For purposes of illumination, means of egress shall consist only of exits and aisles, corridors, passageways, ramps, escalators and lobbies leading to the exits. The one footcandle is for a period of *1 1/2 hours* in the event of failure of the normal lighting. The illumination may decline to .6 footcandle at the end of the 1 1/2 hours.

Exits shall be marked by an approved sign readily visible from any direction of exit access. Access to exits shall be marked by readily visible signs in all cases where the exit or way to reach it is not immediately visible to the occupants. Sign placement shall be such that no point in the exit access is more than 100' from the nearest visible sign. *Every* exit sign shall be *suitably illuminated* by a reliable light source. Externally and internally illuminated signs shall be visible in both the *normal and emergency lighting mode.*

All exit signs for egress elements shall be provided an emergency source of power. Exception: Approved self-luminous signs.

OSHA now requires exit signs in all places of employment in new *and existing buildings.* Today there are some *existing* occupancies which do not have exit signs now required by federal law.

Emergency systems are to maintain illumination or provide power for fire pumps or hospital life-support equipment, in the event of failure of the normal power supply.

Battery-operated unit equipment generally has a test switch to simulate a failure of the normal system and an indicating light that glows brightly while charging and dims when ready.

Lack of testing and maintenance to observe visual signals indicating a malfunction of battery-charging equipment are the main causes of emergency equipment failures.

All boxes, enclosures, transfer switches, generators, and power panels for emergency circuits shall be marked to identify them as a component of the emergency system.

Emergency circuit wiring is not permitted to occupy the same conduit, box, cable, or cabinet with the normal wiring. By keeping the emergency wiring separated a fault on the normal circuit wiring will not affect the emergency circuit except at the transfer equipment enclosure, etc.

Emergency wiring supplying fire equipment is routed so as not to be burned out by a fire in the area they are protecting. Use underground conduits, separate conduit and wire from other circuits, and use fire-resistant insulation cable.

A one hour fire rating of an electrical circuit differs from a one hour fire-resistance rating of a structural member, such as a wall. An electrical conduit and the insulation of the wires must be intact and functioning without any faults, shorts, or open circuits with a one hour fire test. A wall subjected to a one hour fire resistance test must only prevent a fire from passing through or past the wall, without regard to damage to the wall.

Power supply for emergency systems can be one or more of the following:

1. A storage battery, or batteries, provided with an automatic charging means.
2. A prime mover driven generator with capacity to carry the maximum load connected.
3. Two services which may be either underground or overhead, separately electrically and physically.
4. Uninterruptible power supplies (UPS).
5. Unit equipment. Unit equipment is intended to provide illumination for the area where it is installed. It has a rechargeable battery, battery charging means, with two lights. It is required to be connected to the branch circuit supplying the corridor where it is located. In the event of a loss of normal lighting, the unit would automatically energize the unit lamps and provide lighting. The branch circuit feeding the unit must be identified in the panel as such.

UPS power is several times more expensive than emergency power, but saves the product and prevents process interruption. Emergency power is less expensive, but product or process losses may exceed the initial savings in the chosen supply system.

EMERGENCY LIGHTING CIRCUITS

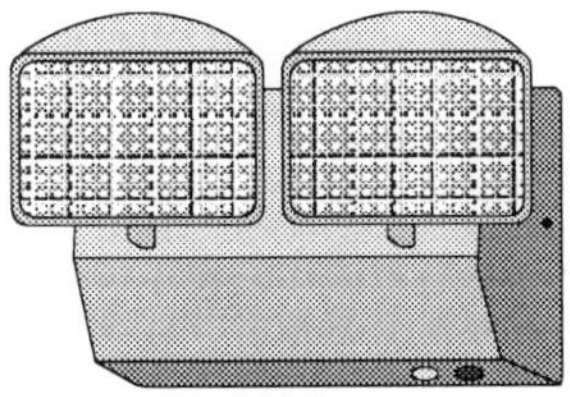

Where emergency lighting is provided by a prime mover driven generator, a delay of not more than 10 seconds shall be permitted. 10 seconds seems like a long time when you're in darkness.

No other lamps or appliances shall be connected to emergency lighting circuits.

Emergency lighting shall include all required means of egress lighting, exit sign lighting, and all other lights needed to provide the required illumination.

Emergency lighting shall be installed so that the failure of any individual bulb will not leave total darkness in that space.

Where high-intensity discharge lighting take time to light after being energized, the emergency lighting system shall be required until normal illumination has been restored. This may require a photo-eye, etc.

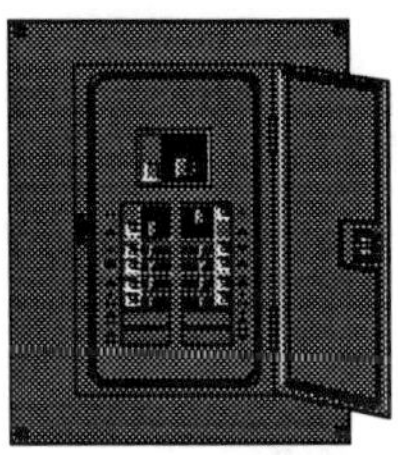

Branch circuits feeding emergency lighting units must be clearly identified at the panelboard.

The lighting unit must be connected ahead of switches so that only an interruption of power at the panelboard will cause the emergency lighting unit to come on.

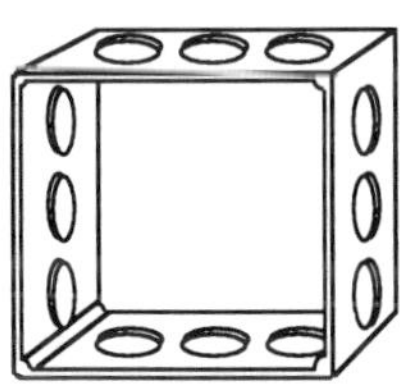

All boxes and enclosures for an emergency system must be marked to indicate they are in the emergency circuit or system.

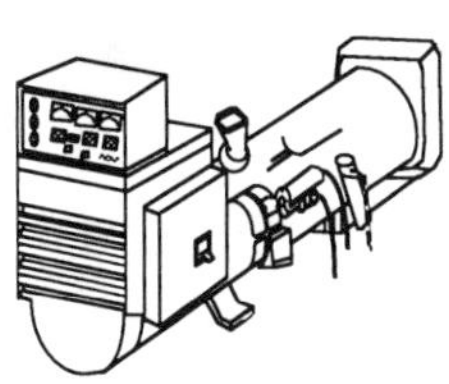

The difference between an emergency system and a legally required standby system is the degree of immediate hazard to life. An emergency system requires the power to be restored within 10 seconds. A legally required standby systems allows 60 seconds.

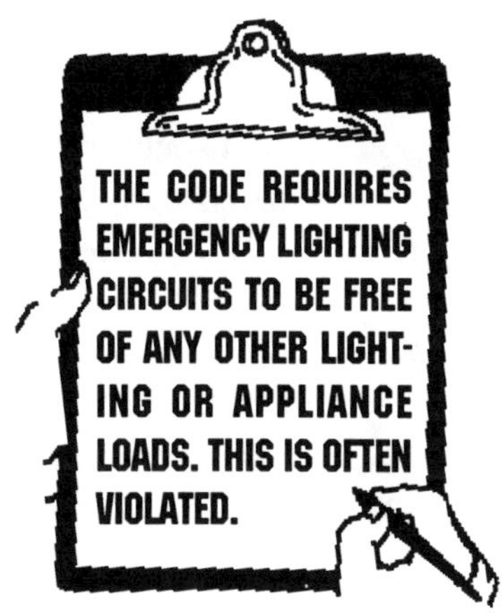

EMERGENCY SYSTEMS (restore power within 10 seconds)

Essential for safety to human life.

The inspector is required by Code to conduct or witness a test of the entire emergency system following completion of installation, and periodically afterwards.

Legally Required Standby Systems (restore power within 60 seconds)

Essential for heating, cooling, refrigeration, sewerage disposal, etc.

The inspector is required by Code to conduct or witness a test of the entire emergency system following completion of installation, NOT thereafter.

OPTIONAL Standby Systems (no time limit on restoring power)

Useful in preventing economic loss, used when safety is NOT a concern.

No acceptance test is required by the inspector. Although the equipment is requested by the owner of the building and is not a Code requirement the inspector must still approve and inspect the installation as a misapplication could cause a safety hazard.

Most inspectors will insist on emergency system equipment to be *listed*. Legally required standby and optional standby systems can be *approved* for use by the inspector.

Both legally required standby and optional standby wiring can be installed within a common enclosure with other wiring.

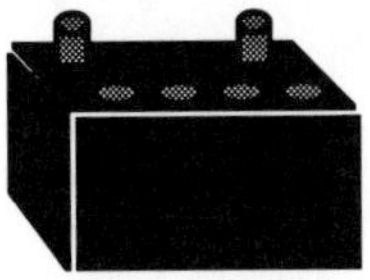

A battery is the most dependable source available for emergency power and can be one of the most versatile when applied with other devices.

The required backup time for batteries for emergency lighting is 90 minutes and the transfer must occur within 10 seconds.

When used for standby service, a double-throw manual or automatic switch should be provided. In areas such as hospitals, theaters, and other buildings where an electrical failure could prove hazardous to the people within, the switch should be automatic so as to supply a certain amount of lighting should the normal power be interrupted.

The location of storage batteries is very important to the safety of life and property, and the electrical inspector must make certain that all storage batteries are located in areas provided with sufficient diffusion and ventilation of gases from the battery to prevent an explosion.

Regular tests should be based on manufacturer's recommended procedure to establish program and adjust from experience. The coordination of overcurrent devices with transfer switch ratings, under fault conditions, is one of the most important aspects of maintaining the integrity and reliability necessary in the operation of an emergency power system.

PREVENTIVE MAINTENANCE CAN BE THOUGHT OF AS A SCIENCE OF ANTICIPATION AND PREDICTION OF FAILURES.

Evaluating the performance of an emergency generator under fault conditions is a critical concern. In most cases, emergency power sources do NOT produce as much fault current as the normal source. Test information on generator short-circuit characteristics may be necessary from the manfacturer.

A large industrial facility in some cases will back up an electrically driven fire pump with a mechanically driven fire pump with a diesel drive.

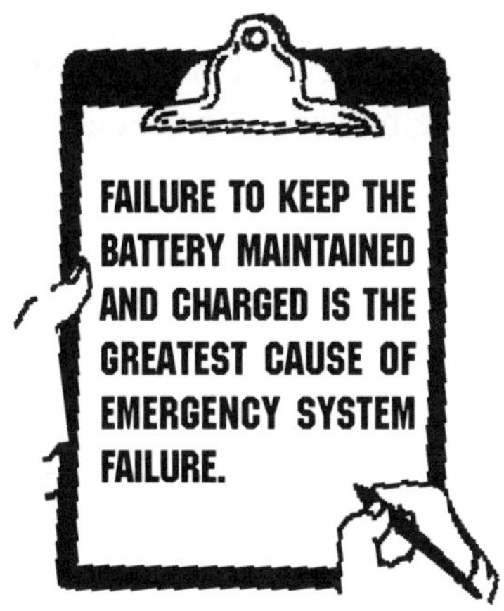

HOUSEHOLD SMOKE DETECTORS

Smoke detectors shall be installed outside of each separate sleeping area in the immediate vicinity of the bedrooms and on each additional story of the family living unit, including basements and excluding crawl spaces and unfinished attics. In new construction, a smoke detector also shall be installed in each sleeping room.

Smoke detectors shall be located on or near the ceiling. Smoke detectors in rooms with ceiling slopes greater than one foot rise per 8 feet horizontally shall be located at the high side of the room.

A smoke detector installed in a stairwell shall be located as to assure that smoke rising in the stairwell cannot be prevented from reaching the detector by an intervening door or obstruction.

The location of the required smoke detectors does not provide adequate warning for the occupants from a fire starting within their bedrooms, nor do the required smoke detectors provide reliable early warning protection for those areas separated by a door. For these reasons, it is recommended that the household consider the use of additional smoke detectors for increased protection.

Tests or inspections, as recommended by the manufacturer, shall be made by the householder not less than once a month for other than battery-powered detectors and not less than once a week for battery-powered detectors.

Fire alarm systems installed in buildings must meet the requirements of NFPA Code 72, and all equipment should be listed.

The electrical inspector should obtain a copy of the manufacturer's specifications and installation instructions and then make certain that the system is actually installed according to these specifications. Upon completion the system should be tested by or in the presence of the electrical inspector.

CHAPTER 8

EXAM

EMERGENCY SYSTEMS - EXAM 1

1. When a diesel engine is used as the prime mover of a generator to supply emergency power, how much site fuel is required?

(a) one-half hour of fuel supply **(b) one hour of fuel supply**
(c) two hours of fuel supply **(d) three hours of fuel supply**

2. Optional standby systems are typically installed to provide an alternate source of electric power for such loads as ____ systems.

I. communications II. data processing III. refrigeration

(a) III only (b) I and III only (c) II and III only (d) I, II and III

3. The branch circuit overcurrent devices in emergency circuits shall be ____.

(a) of the reset type only **(b) a slow-blow type**
(c) accessible to only authorized personnel **(d) painted yellow**

4. Emergency lighting, emergency power or both, in a building or group of buildings will be available within the time required for the application, but not to exceed one of the following:

(a) 5 seconds (b) 10 seconds (c) 30 seconds (d) 60 seconds

5. A storage battery of suitable rating and capacity to supply and maintain at not less than ____ of the system voltage the total load of the circuits supplying legally required standby power for a period of at least 1 1/2 hours.

(a) 100% (b) 75% (c) 50% (d) 87 1/2%

6. The alternate source for emergency systems ____ be required to have ground fault protection of the equipment with automatic disconnecting means.

I. shall II. shall not

(a) I only (b) II only (c) either I or II (d) neither I nor II

7. A storage battery supplying emergency lighting and power shall maintain not less than 87 1/2 percent of full voltage at total load for a period of at least ____.

(a) 2 hours (b) 1 1/2 hours (c) 1 hour (d) 1/4 hour

8. Audible and visual signal devices shall be provided, where practicable to indicate ____.

I. derangement of emergency source
II. that the battery charger is not functioning
III. that the battery is carrying the load

(a) I only (b) II and III only (c) I and II only (d) I, II and III

9. The prime mover of an emergency generator set ____.

I. must be provided with an automatic means for starting
II. must be provided with an automatic means of transferring from one fuel supply to another, where dual supplies are used
III. must have an on-site fuel supply sufficient to operate the prime mover at full demand for 2 hours

(a) I only (b) II only (c) III only (d) I, II and III

10. For emergency systems, the authority having jurisdiction shall conduct or witness a test on the complete system upon installation and periodically afterward. A ____ shall be kept of such tests.

(a) report (b) log (c) written record (d) chart

11. Emergency illumination shall include all required means of ____ and all other lights specified as necessary to provide required illumination.

I. illuminated exit signs II. egress lighting III. show window lighting

(a) I only (b) II only (c) I and II only (d) I, II and III

12. For a legally required standby system, power will be available for the application within ____.

(a) 30 seconds (b) 10 seconds (c) 2 minutes (d) one minute

CHAPTER 9

SIGNS AND HAZARDOUS LOCATIONS

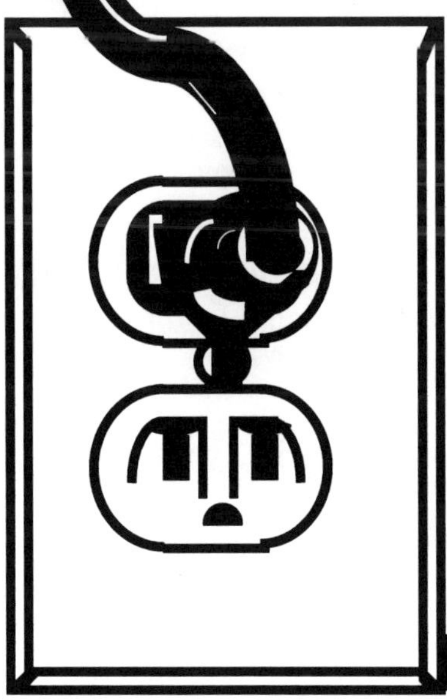

S
I
G
N
S

An electric sign is a fixed, stationary, or portable self-contained, electrically illuminated utilization equipment with words or symbols designed to convey information or attract attention.

A portable sign shall have a weatherproof receptacle and attachment plug with one pole for grounding.

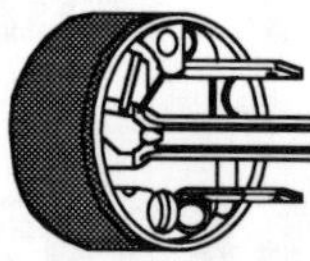

Grounding pin is longer

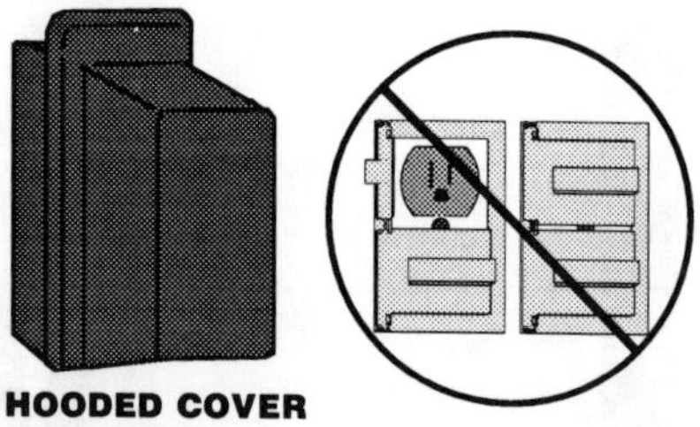

HOODED COVER IS REQUIRED

Cords shall be of the hard service or junior hard service type 3-conductor with one conductor grounded.

Outdoor portable signs shall be provided with *factory-installed* **GROUND-FAULT CIRCUIT-INTERRUPTER** for the protection of personnel. It shall be an integral part of the attachment plug or located in the supply cord within 12" of the attachment plug.

Code section 90-7 states that factory-installed internal wiring or the construction of the equipment does *not need to be inspected* at the time of installation of the equipment, except to inspect for alterations or damage. If the sign has been **LISTED** by a qualified electrical testing laboratory it meets the Code requirements. UL inspections centers are located in 195 cities in the United States and overseas.

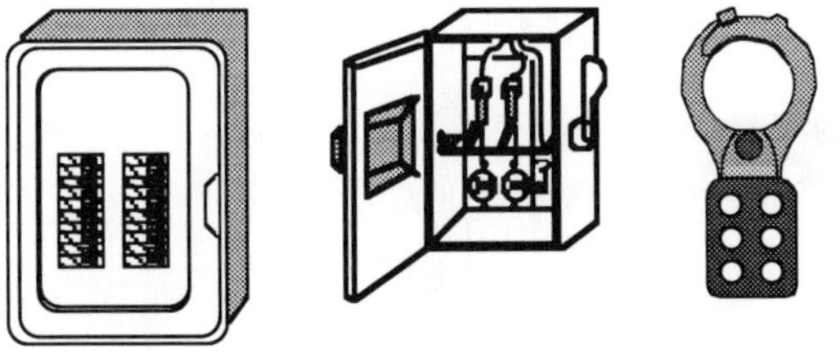

Signs require a disconnect within sight of the sign. If the disconnect is out of sight of the sign it shall be *capable* of being **LOCKED** in the open position.

For a cord connected sign the attachment plug can serve as a disconnect.

Signs in wet locations shall be weatherproof and have at least one **DRAIN HOLE** 1/4" to 1/2" diameter at every low point.

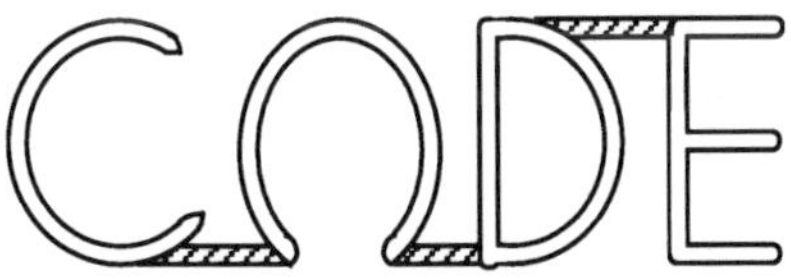

Neon tubing accessible to pedestrians shall be protected from physical damage.

Sign ballasts and transformers shall be accessible and **SECURELY FASTENED IN PLACE**.

Transformers installed in an attic or soffit must have a 2' x 3' access door and a passage way of at least 3' high and 2' wide with a **PERMANENT** walkway at least 12" wide.

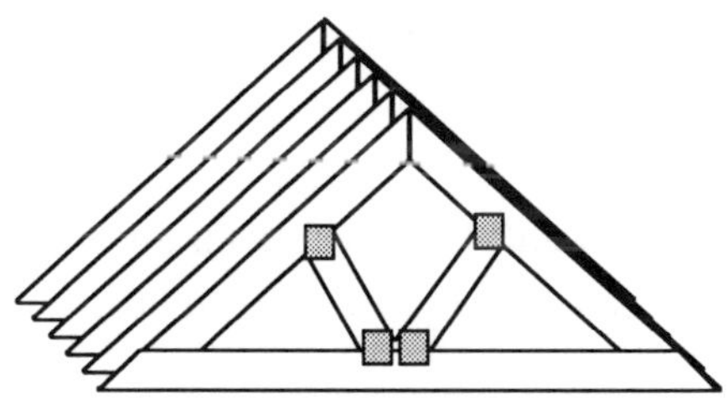

A metal pole can be used as a raceway for conductors supplying a sign.

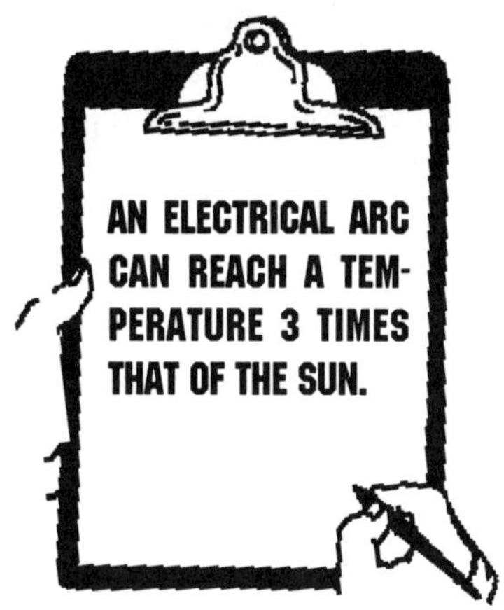

NEON LIGHTING

As simple as a neon tube is, it must be considered a very dangerous piece of equipment.

For neon sign work a step-up transformer is required. The output to the neon sign is generally 15,000 volts.

Ordinary transformers that possess good regulation are not used in neon work. With an ordinary transformer the voltage across the secondary is maintained almost constant regardless of the current being drawn from the secondary.

In neon tube work good regulation of a transformer is not desirable. This comes from the fact that the higher the current through the tube the lower the resistance becomes. If a neon tube were connected across an ordinary transformer, the current would increase. As it increased, the resistance of the gas would decrease, and, as a result, still more current would flow. In a short time, the current would be so high that either the transformer would burn out or the tube would be destroyed.

What is used for neon work is a transformer having *poor regulation*. A neon transformer is built in such a way that as more current is drawn from the secondary, the secondary voltage goes down. This transformer soon reaches a condition of equilibrium, since if too great a current is drawn, the voltage goes down almost to zero, and the increased current can no longer flow.

When no current flows (open-circuit condition), the full 15,000 volts appears in the secondary winding. If now more and more current is taken from the secondary, the voltage decreases. At about 25 milliamperes, the working current for which the transformer is designed to operate, the voltage has dropped to a much lower value. If the two terminals of the secondary (15,000 volts) are connected together (dead short), the maximum current that can flow is 30 milliamperes. The secondary of a neon sign transformer can be short-circuited without harm to the transformer.

Many neon sign manufacturers have claimed that it is injurious to operate a neon transformer for any length of time on an *open* circuit; that is without any load connected to its secondary terminals.

The precautions which must be taken involve the thorough insulation of every metallic part in the high-voltage circuit so that it may not come in contact with anything else.

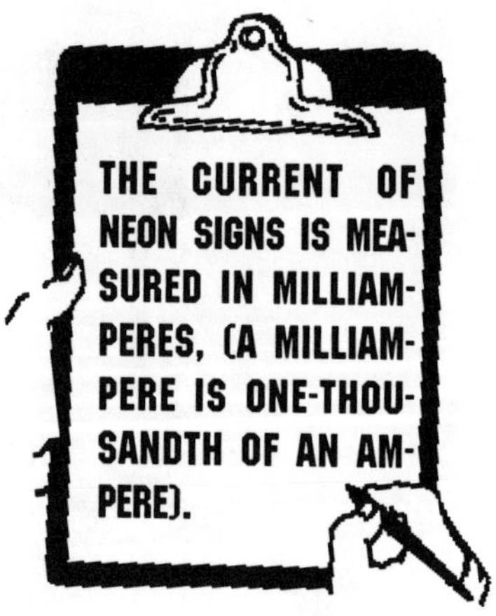

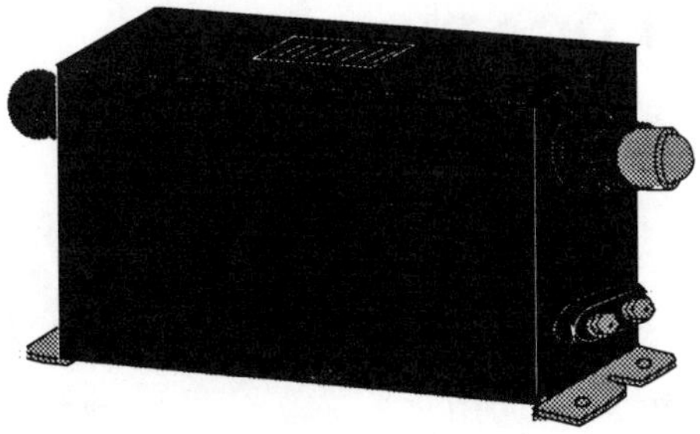

For each size transformer, there is a definite limit to the number of feet of tubing which it can handle.

Even though the high-voltage cable insulation is designed to withstand the full high voltage, if cables from opposite sides of the transformer come into contact, the strain on the insulation may be severe enough to cause eventual breakdown. In this case, each cable should be protected by a *glass tube*, through which it is run throughout its entire exposed length.

UL requirements for wiring of electric signs require that each sign of the gas tube type shall be so wired that not more than one transformer will be dependent upon a single automatic overload protective device unless the combined load is less than 1650 va. Thus, four 400 va transformers may be grouped on a single circuit, but five 400 va transformers would require two circuits.

UL states except at transformer connections, insulated conductors shall not be spliced.

The inspection of the wiring to a neon sign, I feel, is a very critical inspection as many fires have been attributed to the careless installation wiring.

A potential safety hazard may occur from electrostatically induced currents.

Make sure exposed ends of wires are clean. Twist wires together at least three times. Fold splice back and trim, leaving 1/2" maximum lead.

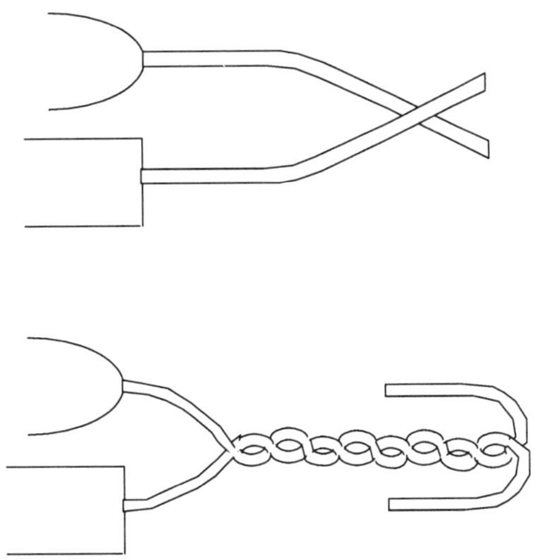

HAZARDOUS (*CLASSIFIED*) LOCATIONS

Most electrical inspectors are reluctant to assume the responsibility for classifying an area or process, preferring to leave that to some other authority, such as a building official or fire marshall. The inspectors wait until after classification is established, and then step in to enforce Code requirements.

The Code becomes more complex for hazardous locations and must be applied with an understanding of other items that must be considered in design and installation.

Testing to assure the equipment's suitability for use in a specific atmosphere is essential. Determining the degree of hazard that will be encountered is another major safety consideration that must be considered in the design of the facility.

Code Article 500 is Hazardous (**CLASSIFIED**) Locations. The word classified makes it clear that certain areas are classified as hazardous.

Who classifies the areas? Generally this is a team effort involving the process engineer, the electrical engineer, the project engineer, and the safety specialist.

A description of the area classification must be written into the electrical design specification for the job.

To have a fire, you must have three things present: Fuel, oxygen, and heat. Take away any one of those three and the fire will be extinguished.

CHEMISTRY OF FIRE

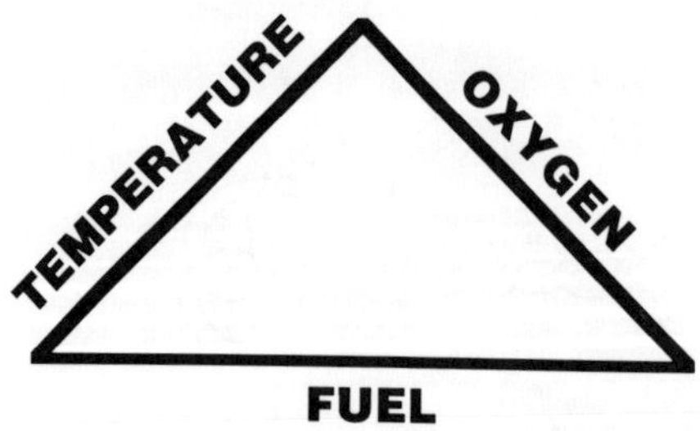

Hazardous locations in industrial plants are involved with a wide variety of flammable gases and vapors and ignitible dusts.

Flammable gases and vapors are, by far, the most common potential sources of an explosion.

Atmospheres containing combustible metal dusts, carbon dust, agricultural dust such as grain, wood, and others such as plastics must be considered to be hazardous.

It is doubtful that it is practical to keep normal atmospheric dust from settling in or on equipment during operation. However, it is absolutely necessary and practical to limit entrance of significant amounts of dust.

Because of the higher level of danger, the installation of electrical equipment in hazardous locations must be done in strict compliance with the instructions given in product standards.

With respect to equipment approval, the OSHA rules on hazardous location equipment must be clearly understood. OSHA rules constitute fereral laws on hazardous locations, which require retroactive rules on all electrical systems, both new and old.

A new **ZONE** classification Article 505 has been added to the Code, however the Code mandates that if Article 505 is used, area classification, wiring, and equipment selection must be under the supervision of a qualified **REGISTERED PROFESSIONAL ENGINEER**.

The weakest link in the total system can undermine the effectiveness of the best individual components.

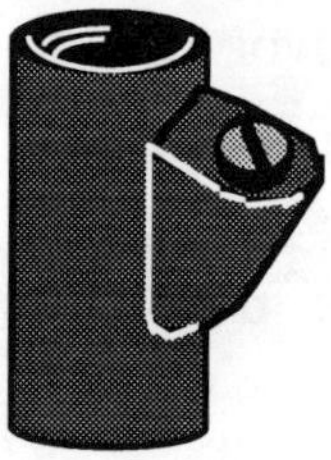

A seal must be vaportight to stop the gases and vapors. The sealing compound must adhere to the fitting and to the conductors.

Sealing compound must expand while it hardens to close all voids. It must develop enough mechanical strength to withstand the forces of explosions.

If the sealing compound is not mixed correctly it will not develop the required mechanical strength.

It is important that the conductors be permanently separated from each other and from the seal fitting walls so the sealing compound will surround each conductor.

When an explosion occurs inside a threaded enclosure, the flame and hot gases create an internal pressure against the cover, thus locking the threads and forcing the gases out through the helical path of threads. By the time the gases reach the outside hazardous atmosphere, they have cooled to a point below igntion temperature.

If threads are damaged, the cover cannot be completely threaded into the body, and **EXPOSED THREADS** warn the inspector that the cover is NOT properly installed. All covers must be tight, even a small opening could allow a flame to pass.

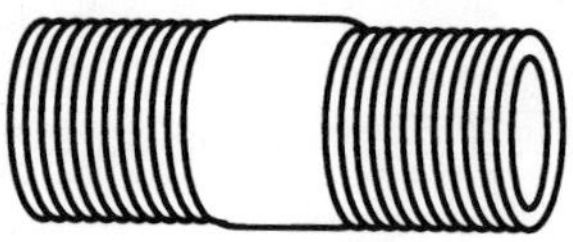

The Code requires the use of rigid metal conduit for explosion-proof equipment with **FIVE FULL THREADS** engaged at entrances to equipment.

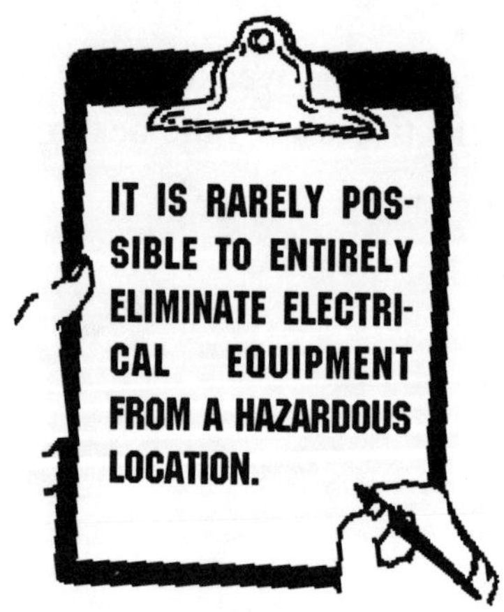

CHAPTER 9
EXAM

SIGNS - EXAM 1

1. Circuits containing electric-discharge lighting transformers exclusively shall not be rated in excess of ____ amps.

(a) 20 (b) 30 (c) 40 (d) 50

2. Electric-discharge tubing operating at more than 7500 volts shall clear the adjacent surface by not less than ____.

(a) 1/4" (b) 1/2" (c) 1" (d) 3"

3. Electric sign enclosures may be constructed of wood if ____.

(a) kept 1" from lampholders
(b) wood is not permitted at all
(c) kept 2" from lampholders
(d) none of these

4. Signs in wet locations shall be weatherproof and ____.

I. have drain holes positioned so there is no external obstructions
II. have at least one drain hole in every low point
III. drain holes shall not be smaller than 1/4"

(a) I and II only (b) II and III only (c) I only (d) I, II and III

5. The cord shall not exceed ____ feet in length for portable or mobile signs in dry locations.

(a) 6 (b) 8 (c) 10 (d) 15

6. Each commercial building and each commercial occupancy accessible to pedestrians shall be provided at an accessible location outside the entrance, with at least one ____ for sign or outline lighting use.

(a) outlet (b) duplex (c) GFCI (d) none required

7. Not more than ___ of high-voltage cable shall be run in a metal raceway from a transformer/power supply to other parts of the sign.

(a) 10' (b) 12' (c) 15' (d) 20'

8. A branch circuit feeding a sign which has a combination of lamps and transformers shall not exceed the rating of ____ amps.

(a) 15 (b) 20 (c) 30 (d) 50

9. Signs operated by electronic or electromechanical controllers located externally to the sign shall have a disconnecting means located ____.

(a) within sight of the sign
(b) within sight from controller location
(c) only in the controller
(d) only externally to the controller

10. Every electric sign of any type, fixed or portable, shall be ____.

(a) listed (b) approved (c) permanently wired (d) electrically isolated

11. Portable outdoor signs shall be protected by ____.

(a) current limiting fuses
(b) instantaneous breakers
(c) GFCI
(d) time delay fuses

12. ____ signs shall be marked with input amperes at full load and input voltage.

(a) All (b) Incandescent (c) Electric-discharge lamp (d) Neon

H A Z A R D O U S E X A M 2

HAZARDOUS LOCATIONS - EXAM 2

1. Enclosures for isolating switches in Class I, Division II locations ____.

(a) may be of general-purpose type
(b) must be explosion-proof type
(c) must have interlocking devices
(d) may not have doors in hazardous area

2. The Code provides that in Class I, Division I locations conduit seals shall be placed not farther from spark-producing devices than ____ inches.

(a) 18 (b) 24 (c) 12 (d) 30

3. Which of the following applies to Class I, Division I locations?

(a) ignitible flammable gases and vapors
(b) grain silos
(c) ignitible fibers or flyings
(d) combustible dust

4. For general wiring in Class I, Division I locations it is permissible to use ____.

(a) rigid metal conduit (b) EMT (c) flexible metal conduit (d) all of these

5. Sealing compound is employed with mineral-insulated cable in a Class I location for the purpose of ____.

(a) preventing passage of gas or vapor
(b) excluding moisture
(c) limiting a possible explosion
(d) preventing escape of powder

6. Locations where combustible dust is normally in heavy concentrations are designated as ____.

(a) Class I, Division II **(b) Class II, Division I**
(c) Class II, Division II **(d) Class III, Division I**

7. What is the minimum burial depth for rigid nonmetallic conduit in a dispensing station Class I, Division I location?

(a) 18" (b) 24" (c) 30" (d) cannot be used in Class I, Division I location

8. Where magnesium, aluminum or aluminum-bronze powders may be present, transformers ____.

(a) must be dust-tight
(b) may be pipe-ventilated
(c) must be approved for Class II, Division I
(d) are not allowed

9. In a Class II location, where electrically-conducting dust is present, flexible connections at motors could be made with ____.

(a) flexible metal conduit
(b) type AC armored cable
(c) hard usage cable
(d) liquid-tight flexible metal conduit with approved fittings

10. Locations where combustible fibers are stored are designated as ____.

(a) Class II, Division II
(b) Class III, Division I
(c) Class III, Division II
(d) non-hazardous

11. In commercial garages, generally the floor area to a height of 18" above grade is designated as ____.

(a) Class I, Division I
(b) Class I, Division II
(c) Class II, Division II
(d) Class II, Division I

12. In normal practice, the NEC discourages the practice of switching the grounded conductor. However, in one of the following cases, the Code makes an exception and actually requires that this be adopted. Select that case.

(a) electrical circuits that lead or pass through wall mounted fixtures mounted at a height of less than 5' 6" and located in pediatric wards
(b) electrical circuits that lead into or pass through wet niche light fixtures located in swimming pools
(c) electrical circuits that lead into or pass through a gasoline dispensing pump
(d) electrical circuits that lead into or pass through paint spray booths

HAZARDOUS EXAM 3

HAZARDOUS LOCATIONS - EXAM 3

1. All of the following motors are permitted in a Class III, Division I area except ____.

(a) totally enclosed pipe ventilated **(b) non-ventilated**
(c) totally enclosed fan cooled **(d) water cooled**

2. An area must be classed as a Class II hazardous location if it contains ____.

(a) combustible dust **(b) ignitable vapors** **(c) flammable gases** **(d) ignitable fibers**

3. All fixed wiring above Class I locations in a repair garage shall be in ____.

I. flexible nonmetallic conduit II. rigid nonmetallic conduit III. TC cable

(a) I only **(b) II only** **(c) I and II only** **(d) I, II and III**

4. Parking garages used for parking or storage and where no repair work is done, open flame, welding, or the use of volatile flammable liquids are ____.

(a) Class I **(b) Class II** **(c) Class III** **(d) not classified**

5. Which of the following would **not** be approved in all Class II locations?

(a) flexible connections **(b) threaded bosses**
(c) dust-tight boxes **(d) EMT**

6. In a Class I, Division II location a conduit passing through into a nonhazardous location, the sealing fitting shall be permitted ____.

(a) no seal required under the conditions **(b) on either side of the boundary**
(c) on both sides of the boundary **(d) at the first fitting**

7. A/an ____ circuit is a circuit in which any spark or thermal effect is incapable of causing ignition of a mixture of flammable or combustible material in air under prescribed test conditions.

(a) low voltage (b) intrinsically safe (c) hazard-proof (d) explosive-proof

8. Storerooms and similar areas adjacent to aircraft hangars but effectively isolated shall be designated ____.

(a) Class I, Division II **(b) Class II, Division I**
(c) Class II, Division II **(d) shall not be classified**

9. For limited flexibility for motor connections in a Class I, Division II location, flexible conduit ____.

(a) must be explosion-proof
(b) must be liquid-tight flexible conduit or equal
(c) may be standard flexible metal conduit
(d) shall not be used

10. In Class I, Division I locations, the Code requires conduit seals adjacent to boxes containing splices if the conduit is equal to or larger than ____.

(a) 3/4" (b) 1 1/2" (c) 1" (d) 2"

11. Lighting fixtures installed over vehicle lanes inside a commercial garage shall be installed a minimum of ____ feet.

(a) 8 (b) 10 (c) 12 (d) 15

12. An approved seal shall be provided in each conduit run entering or leaving a dispenser or any cavities or enclosures in direct communication therewith. The sealing fitting shall be _____ .

(a) concrete-tight
(b) 3/4" minimum thickness
(c) the last fitting after the conduit emerges from the earth or concrete
(d) the first fitting after the conduit emerges from the earth or concrete

CHAPTER 10

MISCELLANEOUS INSPECTIONS

COMMUNICATIONS

Generally telephone systems are installed by the phone company. The electrical contractor usually installs the conduits and outlet boxes, and a pull wire inside the conduit.

Communication conductors must not be installed in any raceway, box, or similar fitting with conductors of light or power circuits, except where separated by a suitable partition.

The inspector is required in commercial buildings such as restaurants, hotels, hospitals, theaters, schools, offices, etc. to evaluate the installation of various sound systems. These systems may include intercommunicating telephone, nurses' call systems, public address, sound recording, and other audible signal equipment and wiring.

Most of the circuits on sound systems will be low voltage except with the power supply.

When communication conductors are installed overhead, proper clearances from power circuits must be maintained as well as clearances over streets, roads, and driveways.

Code Article 820 applies to most community antenna systems using coaxial cable for the distribution of the radio and television frequency signals. When the wiring system employed is other than coaxial, Article 800 of the Code shall apply.

Where practicable, coaxial conductors on poles should be located **BELOW** the light or power conductors and must not be attached to a crossarm that carries light or power conductors.

During any inspection, if certain areas are not clear, the inspector should make notes and perhaps take pictures of the situation for further study before either approving or rejecting a job.

Some jobs have high voltage power circuits to feed gaseous tubing and X-ray equipment.

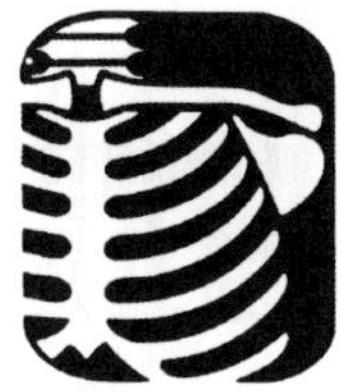

When the inspector encounters projects dealing with the splicing of high voltage conductors, he should make certain that the person performing the splicing is qualified. The splicing of cables is a specialized operation that should be performed by capable workers who have acquired the knowledge of and the ability to perform such operations.

Control wiring and feeder connections on nonportable, electrical driven machines is covered in Code Article 670 Industrial Machinery. The installation of busways is covered in Article 364.

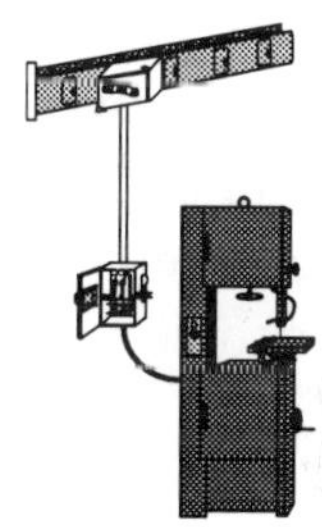

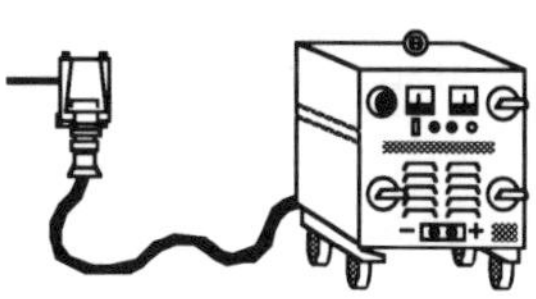

Electric welding equipment is found in Article 630 and is generally treated as a piece of industrial power equipment for which are provided branch circuits adequate for the large amounts of current at a relatively low voltage delivered to the welding electrode. Certain specific conditions, however, apply to circuits feeding AC transformer and DC rectifier arc welders, motor-generator arc welders; resistance welders, etc.

The electrical inspector can benefit by contacting different manufacturers to obtain installation procedures and specifications and build your own library of references.

MISCELLANEOUS INSPECTIONS

When an electrical inspector is required to make an inspection on a crane or hoist, he should become thoroughly familiar with Article 610.

Code Article 620 covers elevators, escalators, moving walks, wheelchair lifts, etc. and in some cases requires a special certified inspector.

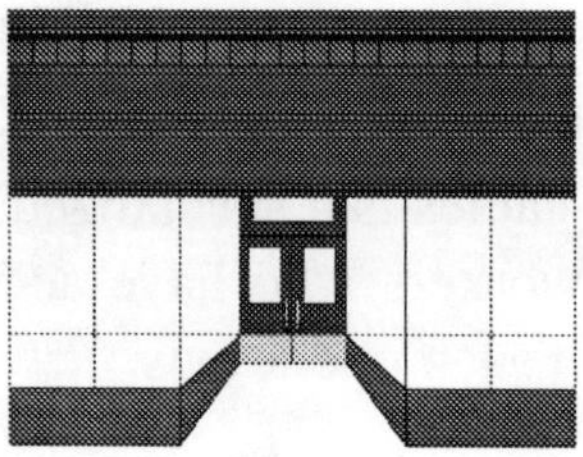

Article 518 covers Places of Assembly which is defined as a building or structure designed or intended for the assembly of 100 or more persons.

Article 520 covers theaters, audience areas of motion picture and television studios, etc. The motion picture studio itself is covered under Article 530, and the motion picture projectors under Article 540.

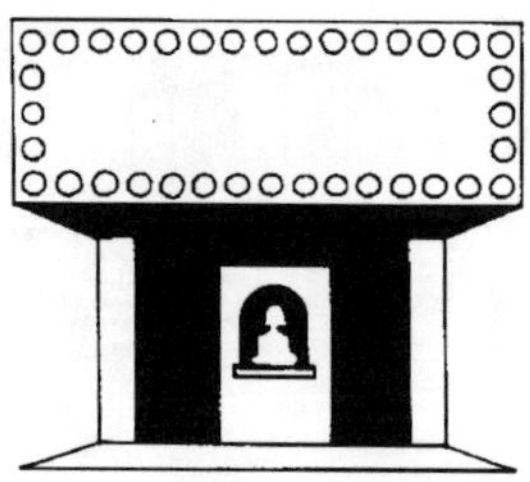

A newer Article 525 covers carnivals, circuses, fairs, etc. areas that really need electrical inspections due to the constant set-ups and break-downs, leads to a degradation of electrical equipment and conductors. Electrical inspectors needed this enforceable set of standards to ensure the public safety for those who use these rides.

Mobile home and travel trailer parks continue to expand requiring more inspections in your jurisdiction. They are covered under Articles 550 and 551.

CHAPTER 10
EXAM

CHAPTER 10 - EXAM 1

1. The minimum size conductor permitted in parallel for elevator lighting is ____, provided the ampacity is equivalent to a #14 wire.

(a) #14 (b) #20 (c) #16 (d) #1/0

2. The neutral of feeders supplying solid-state, 3-phase, 4-wire dimming systems shall be considered a ____ conductor.

(a) current-carrying (b) noncurrent-carrying (c) balanced (d) isolated

3. The Code provides that unshielded lead-in conductors of amateur transmitting stations shall clear the building surface which is wired over by a distance not less than ____ inches.

(a) 1 (b) 2 (c) 3 (d) 4

4. The ampacity for the supply conductors for a resistance welder with a duty cycle of 15% and a primary current of 21 amps is ____ amps.

(a) 9.45 (b) 8.19 (c) 6.72 (d) 5.67

5. The minimum size service for a mobile home in a mobile home park is ____ amps.

(a) 80 (b) 70 (c) 200 (d) 100

6. A motel conference room is designed for the assembly of 100 or more persons. The room is fire-rated construction. One of the following wiring methods shall be required:

(a) Rigid nonmetallic conduit
(b) MI cable
(c) Nonmetallic sheathed cable
(d) Type AC cable

7. Selecting a primary ____ location to achieve the shortest practical primary protector grounding conductor will help limit potential differences between communications circuits and other metallic systems.

(a) limiter (b) protector (c) neutral (d) common

8. Elevators shall have a single means for disconnecting all ungrounded main power supply conductors for each unit; ____.

(a) this does not include the emergency power service if the system is automatic
(b) this does include the emergency power service
(c) this does not include the emergency power service
(d) no elevators are to operate on emergency power systems

9. In communication circuits the bonding together of all separate electrodes ____.

(a) shall not be permitted
(b) shall be permitted with a minimum size jumper #4
(c) shall be permitted with a minimum size jumper #6
(d) shall be permitted with a minimum size jumper #8

10. In a recreational vehicle, which major appliance, other than built-in does the Code consider portable if cord-connected?

I. refrigerator II. clothes washers III. gas range equipment

(a) I only (b) I and II only (c) II only (d) I, II and III

11. All lights and receptacles in dressing rooms of theaters shall be controlled by wall switches installed in the ____.

(a) dressing rooms (b) control room (c) projection room (d) stage office

12. Where practicable, a separation of at least ____ feet shall be maintained between any coaxial cable and lightning conductors.

(a) 2 (b) 2 1/2 (c) 4 (d) 6

CHAPTER 10 - EXAM 2

1. A crane rail when used as a conductor shall be _____.

I. grounded by fittings used for suspension or attachment of the rail
II. grounded by a bonding conductor to a water pipe
III. effectively grounded at the transformer

(a) I only (b) I and II only (c) I and III only (d) II and III only

2. Each receptacle of DC plugging boxes shall be rated at not less than _____ amps.

(a) 15 (b) 20 (c) 25 (d) 30

3. Proscenium side lights shall have the lampholder terminals kept away at least _____ from the metal of the trough.

(a) 1/2" (b) 1" (c) 3" (d) 6"

4. Stage cables used in motion picture studios for stage lighting shall be protected by means of overcurrent devices set at not more than _____ of the values given in the appropriate Code table.

(a) 500% (b) 400% (c) 200% (d) 100%

5. A resistance welder shall have overcurrent primary protection set at not more than _____ percent.

(a) 200 (b) 300 (c) 250 (d) 125

6. The service to a mobile home may consist of _____.

I. 3 - 50a separate services II. one permanently installed feeder

(a) I only (b) II only (c) both I and II (d) neither I nor II

7. The ____ and the bridge frame shall be considered as electrically grounded through the bridge and trolley wheels and its respective tracks unless local conditions, such as paint or other insulating material, prevent reliable metal-to-metal contact.

(a) trolley frame (b) track (c) trolley wheels (d) none of these

8.Only composition or metal-sheathed, porcelain, keyless lampholders equipped with suitable means to guard lamps from physical damage and from film and film scrap shall be used at ____ tables.

I. cutting II. viewing III. patching

(a) I only (b) II only (c) III only (d) I, II and III

9. Conductors supplying outlets for arc or Xenon projectors of the professional type shall not be smaller than ____.

(a) #12 (b) #10 (c) #8 (d) #6

10. The minimum size equipment grounding conductor shall not be required to be larger than ____ copper for wireways not containing power-supply conductors for Audio signaling.

(a) #22 (b) #18 (c) #16 (d) #14

11. Amusement rides shall be maintained not less than ____ feet in any direction from overhead conductors operating at 600 volts or less, except for the supply conductors to the ride.

(a) 8 (b) 10 (c) 12 (d) 15

12. Industrial machinery is defined as ____.

I. a portable machine used to shape or form plastic
II. a power-driven machine not portable by hand

(a) I only (b) II only (c) I and II (d) neither I nor II

CHAPTER 11

FINAL INSPECTION

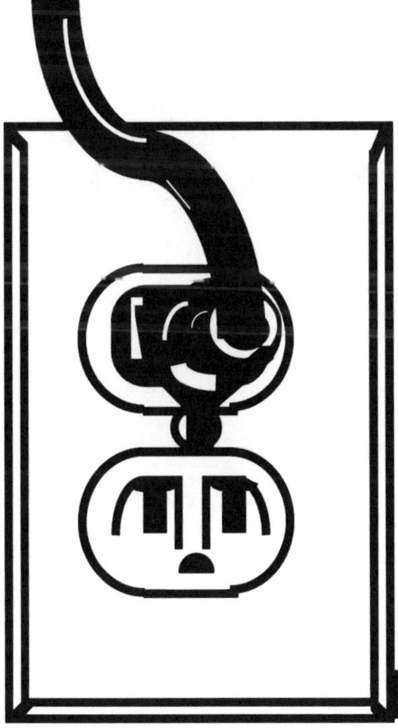

FINAL INSPECTION

A final inspection of the building is required before a CO (certificate of occupancy) can be issued to the owner.

The electrical inspector must be positively certain that the entire electrical installation is in conformance with the requirements of the Code before granting a final sign off.

The final sign off of the building is generally from the electrical inspector after he has been informed by all the other inspectors that all corrections have been made. The building cannot function without electrical power and the electrical inspector has the final say.

As you gain experience you'll soon be able to format your own personal check list.

It would be impossible to have only one check list as you will be called to make final inspections for all types of buildings, add-on's, signs, etc. I will include some you'll want to make note of.

•Test all circuits for proper operation and the entire system to be free of grounds and shorts.

•Proper polarity of receptacles and fixtures.

•Head room clearance and lighting at electrical equipment.

•Service conductors must be of a type to withstand exposure to atmospheric and other conditions of use without detrimental leakage of current.

•Check hot water pipes to be sure cables are not crossing on them.

•Check fixtures for correct wattage of bulbs.

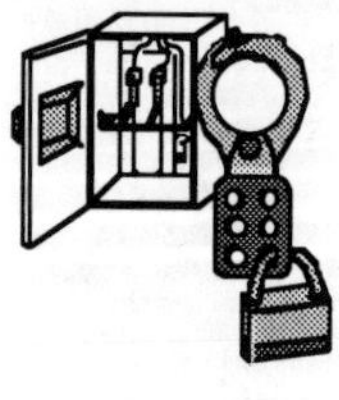

•On commercial and industrial buildings the inspector should check all fusible safety switches to see if the handles will accept heavy duty type padlocks for lock-out purposes.

•The panel directories should be neatly typed indicating the function of each circuit breaker.

•When inspecting electrical heating cable installations, carefully check to make sure that the cables have been installed in their complete lengths as supplied by the factory. Under no circumstances should the cables be shortened.

•Check terminal lugs for cutting and removing of some of the strands of the conductors. This lowers the conductors ampacity.

When applying the variables to circuit designing often it is required to use a larger conductor for voltage drop, ampacity corrections, cost, etc. The question that comes up is how can you properly install a #4 conductor to a 20 amp circuit breaker? The terminal lug isn't big enough.

WIRE RANGE	
Maximum	Minimum
#2	#14
#1/0	#14
#4/0	#6
#250	#6
#350	#6
#500	#3/0
#750	#250
#1000	#500

UL listed splicer/reducer lugs are available for this purpose.

•Proper working clearances about electrical equipment.

•All devices in place with proper cover plates installed.

•Electrical equipment shall be plumb and level in a workmanlike manner.

•Terminals for more than one lug shall be so identified.

•Circuit breakers and fuses properly sized to the conductors.

•All equipment exposed to the weather is approved for the condition.

•GFCI protection installed where required.

•Inhibitor applied to aluminum conductors where required.

•Terminals torqued to proper requirements.

•Effective grounding system installed.

•Test all emergency systems.

INSPECTING THE PANELBOARD

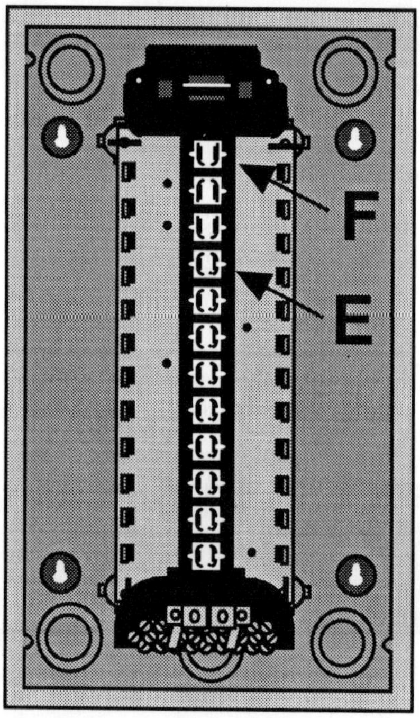

Class CTL is a UL designation for *circuit limiting* which the Code requires not more than 42 overcurrent devices.

The slots in the panelboard have different configurations to limit the total number of spaces to no more than 42.

Each "F" slot configuration will accept only one pole. Each "E" slot configuration will accept two poles.

Not all panelboards are designed or listed for tandem breakers.

Panelboards which are designed to accept tandem breakers are designed so the mounting cam on the tandem breaker can only be installed in a position that has a mounting pan rail slot which will accept the cam. This prevents the panelboard from having over 42 overcurrent devices.

TANDEM BREAKER

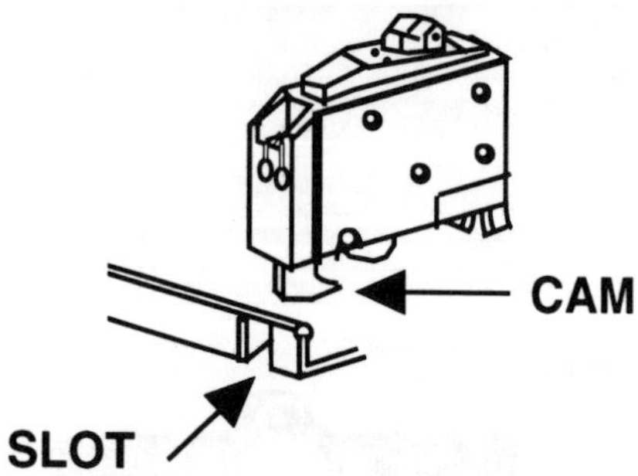

A *violation* of the UL listing and the Code occurs when the cam is broken off so the tandem breaker can be installed into a space that was not designed to receive it.

•The wiring in a panelboard should be formed so you can connect an ammeter.

SUMMARY

This book was NOT written as a step by step guide of how to make a total electrical inspection, but rather as a guide to make certain the electrical inspector is not overlooking the **CRITICAL AREAS** that have started fires and caused accidents.

The point this workbook has tried to make is, let's inspect the building for the hazards that are causing damage to the building and injuries to people, so if we have our day in court we can state we have carefully examined the **HAZARDS** within the building.

If a Code violation should occur, a competent inspector will explain the Code section and the intent of the rule to the electrician. A competent and knowledgeable inspector will work with the electrician to create a harmonious working environment.

If you are the competent inspector then you are the respected one who takes the time to give your expert opinion and to impart your knowledge with integrity and provide instruction to help the electrical industry become more educated in the rules of the Code.

Inspection is a very demanding career, and the technical expertise required to handle the job means that individuals today tend to be more educated than before. Education is the key.

Thc world works and lives at the end of a wire. The world needs people to inspect the wire *correctly.*

CHAPTER 12

FINAL INSPECTORS EXAM

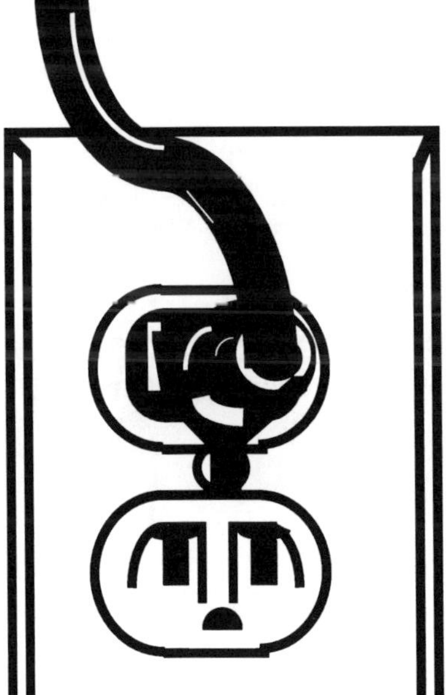

50 OPEN BOOK CODE QUESTIONS

THE FINAL ELECTRICAL INSPECTOR EXAM IS A 50 QUESTION OPEN BOOK EXAM WITH A 2 HOUR TIME LIMIT

1. Which of the following lighting fixture installations is covered by the NEC?

(a) Light fixtures in a utility owned transformer vault located within a generating plant.
(b) Fixtures on poles provided for outdoor lighting within the fenced area of a utility owned substation.
(c) Fixtures on poles along a private farm road where the utility has established rights of easement.
(d) Fixtures on poles provided for outdoor lighting of the employee parking lot of a utility owned office building.

LIST CODE SECTION ____________________.

2. "A load where the wave shape of the steady state current does not follow the wave shape of the applied voltage" is the NEC definition for ____ load.

(a) neutral (b) nondemand (c) nonlinear (d) unbalanced

LIST CODE SECTION ____________________.

3. An electronically actuated fuse generally consists of all of the following EXCEPT ____.

(a) a control module that provides current sensing
(b) electronically derived time-current characteristics
(c) an interrupting module that interrupts current when an overcurrent occurs
(d) a thermally sensitive part that is heated and severed by the passage of overcurrent through it

LIST CODE SECTION ____________________.

4. Lighting controlled ONLY by automatic means is specifically prohibited for ____.

(a) hallways (b) stairways (c) storage rooms (d) electrical equipment rooms

LIST CODE SECTION ____________________.

5. An underground pull box used for circuits of over 600 volts shall have the cover locked, bolted or screwed on, or the cover is required to weigh over ____ pounds.

(a) 25 (b) 50 (c) 75 (d) 100

LIST CODE SECTION ____________________.

6. Given: For an outdoor electrical installation of over 600 volts, the plan shows a fence installed to deter access by unqualified persons. The plan shows a one foot extension at the top of the fence with three barbed wire strands.

The height of the fence below the one foot extension shall be at least ____ feet high.

(a) 5 (b) 6 (c) 7 (d) 8

LIST CODE SECTION ____________________.

7. Given: On a circuit where a grounding means does not exist, a nongrounding-type receptacle is replaced with a ground fault circuit-interrupter-type (GFCI) receptacle which supplies no other receptacles.

This new GFCI receptacle shall be marked ____.

(a) "Not Grounded" **(b) "GFCI Protected"**
(c) "No Equipment Ground" **(d) "No Grounded Conductor"**

LIST CODE SECTION ____________________.

8. Ground fault circuit-interrupter (GFCI) protection is required in all of the following locations EXCEPT ____.

(a) kitchen receptacles in an office building lunchroom installed within 6' of the sink
(b) kitchen receptacles in a dwelling installed to serve countertop surfaces 10' away from the sink
(c) receptacles in an office building restroom which has only a basin and toilet
(d) a receptacle provided for servicing a rooftop air conditioning unit on the roof of a warehouse

LIST CODE SECTION ____________________.

9. In a dwelling unit, a receptacle outlet installed in the floor may be counted as a required outlet for receptacle wall spacing requirements if it is located a maximum of ____ inches from the wall.

(a) 6 (b) 12 (c) 18 (d) 24

LIST CODE SECTION ____________________.

10. In the kitchen of a dwelling unit, a receptacle shall NOT be installed below a countertop which extends more than ____ inches beyond its support base.

(a) 6 (b) 9 (c) 12 (d) 14

LIST CODE SECTION ____________________.

11. For dwelling units, all of the following are true EXCEPT ____.

(a) outdoor outlets are permitted to be supplied through the small appliance branch circuits
(b) the outlet for kitchen refrigeration equipment may be supplied by an individual 15 amp branch circuit
(c) bathroom receptacles shall be supplied by a 20 amp branch circuit which shall have no other outlets
(d) the clothes washer shall be supplied by a 20 amp branch circuit and outlets outside the laundry area are NOT permitted on this circuit

LIST CODE SECTION ____________________.

12. Given: In a new house, a clothes dryer will be plug-connected to a 120/240 volt circuit with all conductors #10 copper run in 10' of flexible metal conduit.

The dryer cord will have four conductors with a grounding conducting bonding the metal frame of the dryer to the fixed grounding contact of the attachment plug.

The grounding terminal of the dryer receptacle, mounted in a metal box, shall be permitted to be grounded by ____.

(a) the grounded conductor of the circuit supplying the dryer
(b) a bare equipment grounding conductor run in flexible metal conduit
(c) the flexible metal conduit and connector by connecting an equipment grounding jumper from the box to the grounding terminal
(d) an insulated #10 copper wire, run outside the conduit and connected to interior metal water piping within 5' of where the piping enters the building

LIST CODE SECTION ____________________.

13. Given: A pull box is installed in a run of 1" rigid metal conduit containing motor circuit conductors of 480 volts.
For connection to the pull box, the conduit is threaded and two locknuts are used, one inside and one outside the box.
The pull box has concentric knockouts which provide for conduit sizes up to 1 1/2".
The knockouts have been tested to ensure electrical continuity and the box is listed for the use.

For this installation, which of the following statements is a correct code provision for bonding the raceways and pull box?

(a) A single locknut on the outside and a bushing on the inside.
(b) Two locknuts (one inside and one outside the box) is an acceptable method of bonding.
(c) A properly sized bonding jumper is required to be installed between the raceway and box with sheet metal screws.
(d) Threadless connectors with a shoulder that seats firmly against the box with a locknut inside the box are required.

LIST CODE SECTION ____________________.

14. Metal water piping, which is not effectively grounded but is located in the area of a separately derived system, shall be ____.

(a) bonded to a supplemental grounding electrode
(b) used as the grounding electrode of the separately derived system
(c) bonded to the grounded conductor of the separately derived system
(d) bonded to the nearest available effectively grounded structural metal member of the structure

LIST CODE SECTION ____________________.

15. For temporary wiring on a construction site, feeders shall originate in an approved distribution center and ____.

I. conductors shall be permitted within cable assemblies
II. conductors shall be permitted to be multiconductor cord or cable if hard usage type
III. conductors shall be permitted to be multiconductor cord or cable if extra-hard usage type

(a) I only (b) I & II only (c) I & III only (d) I, II or III

LIST CODE SECTION ____________________.

16. Given: A receptacle in the hallway of a dwelling is available to provide power for electrical hand tools used by personnel during the remodeling of the dwelling.
This receptacle is part of the permanent wiring of the dwelling and is NOT affected by the remodeling.

For use by construction personnel for temporary power purposes, this receptacle ____.

(a) is NOT permitted to be used
(b) shall have GFCI protection provided
(c) is permitted to be used without GFCI protection
(d) is permitted to be used without GFCI protection only by qualified personnel

LIST CODE SECTION ____________________.

17. Given: In an industrial building where only qualified persons will service the installation, a run of 3" electrical metallic tubing (EMT) is terminated at a cable tray. The EMT, at its termination, is supported by and is attached to the cable tray with a listed clamp designed for support.

The EMT in this installation ____.

(a) is NOT permitted to be supported by the cable tray
(b) requires NO nearby support in addition to the cable tray
(c) requires support within 3' of the end of the cable tray
(d) requires support within 5' of the end of the cable tray

LIST CODE SECTION ____________________.

18. Given: An unbroken length of rigid metal conduit shall be permitted to support a box used for fixture support where several conditions are met.

One of the conditions for this installation is that the maximum weight supported by any single conduit does not exceed ____ pounds.

(a) 10 (b) 20 (c) 35 (d) 50

LIST CODE SECTION ____________________.

19. Nonmetallic auxiliary gutters shall have expansion fittings installed where expected length change due to expansion and contraction caused by temperature change is more than ____ inch.

(a) .25 (b) .50 (c) .75 (d) 1.00

LIST CODE SECTION ___________________.

20. All of the following code provisions are correct for track lighting, EXCEPT ____.

(a) heavy duty track lighting is identified for use exceeding 20 amperes
(b) each fitting attached to heavy duty track lighting shall have individual overcurrent protection
(c) for load calculations, a maximum of 2' of track lighting or fraction thereof shall be considered 180va
(d) track lighting is specifically prohibited within a zone 3' horizontally and 8' vertically from the top of a bathtub rim

LIST CODE SECTION ___________________.

21. Given: A duplex receptacle is installed in a wet location and has a weatherproof cover with one lid to cover both receptacles. When the lid is opened and an attachment plug cap is inscrtcd, thc face of the receptacle remains exposed.

This receptacle cover would comply with the code ONLY if the receptacle is ____.

(a) a ground fault circuit-interrupter-type
(b) used for portable tools or equipment normally connected to the outlet only when attended
(c) used for two attachment plug caps that will be connected continuously, such as for two vending machines
(d) installed in a flush-mounted outlet box and the faceplate assembly provides a watertight connection between the cover and the wall surface

LIST CODE SECTION ___________________.

22. Lighting systems operating at 30 volts or less shall be supplied from a maximum ____ ampere branch circuit.

(a) 15 (b) 20 (c) 25 (d) 30

LIST CODE SECTION ________________.

23. The disconnecting means for a ride at a traveling carnival shall be within sight and within ____ feet of the operator's station.

(a) 6 (b) 8 (c) 10 (d) 15

LIST CODE SECTION ________________.

24. Where cord and plug connected electric vehicle supply equipment is used, the ground fault circuit-interrupter protection for personnel, where not an integral part of the attachment plug, shall be located in the power supply cable not more than ____ inches from the attachment plug.

(a) 12 (b) 18 (c) 24 (d) 30

LIST CODE SECTION ________________.

25. An outdoor swimming pool constructed on or above the ground is defined as a permanently installed pool if it is capable of holding water in a depth greater than ____ inches.

(a) 36 (b) 42 (c) 48 (d) 60

LIST CODE SECTION ________________.

26. The minimum headroom required in front of a motor control center shall be ___ feet.

(a) 6 (b) 6 1/4 (c) 6 1/2 (d) 7

LIST CODE SECTION ________________.

27. Given: THHN conductors are connected to a 125 amp, 2 pole circuit breaker which is listed and marked at 60/75° C. The conductors feed equipment which has terminals listed and marked at 60° C.

The maximum ampacity of the conductors used in this installation shall be determined from the ___ column of NEC Table 310-16.

(a) 60° C (b) 75° C (c) 85° C (d) 90° C

LIST CODE SECTION ________________.

28. Branch circuits are ___ in accordance with the maximum permitted ampere rating or setting of the overcurrent device used on the circuit.

(a) rated (b) listed (c) labeled (d) classified

LIST CODE SECTION ________________.

29. Given: In a new ten unit apartment building, electrical maintenance will be under supervision of, and provided by the building manager occupying one of the apartments.

The branch circuit for the central alarm system, ___ supplied from the panel in the manager's apartment.

(a) is required to be
(b) is prohibited from being
(c) is allowed by special permission to be
(d) if limited to a 15 amp circuit, is allowed to be

LIST CODE SECTION ________________.

30. A factory installed duplex receptacle in a baseboard heater, where the heater is to be permanently installed in a commercial building, is ___.

(a) prohibited by the code
(b) allowed only when the receptacle is factory connected to the heater circuit
(c) not allowed to be used as the required receptacle outlet for flexible cords with attachment plugs, when wired on a separate circuit from the heater circuit
(d) allowed to be used in lieu of the required receptacle outlet for flexible cords with attachment plugs, when wired on a separate circuit from the heater circuit

LIST CODE SECTION ________________.

FINAL EXAM

F I N A L E X A M

31. When heating and air-conditioning equipment is located on a rooftop, a single phase, 15 or 20 ampere receptacle outlet shall be located on the same level within ___ feet from the equipment.

(a) 12 (b) 15 (c) 25 (d) 75

LIST CODE SECTION ________________.

32. In a dwelling unit garage, a non-GFCI protected receptacle, installed in a dedicated space for a single, cord and plug connected stationary appliance, shall be ___.

(a) a single receptacle
(b) prohibited by the code
(c) marked "not GFCI protected"
(d) installed at least 6' 6" above the floor of the garage

LIST CODE SECTION ________________.

33. Given: A metal underground water pipe is used as a grounding electrode and used to bond other electrodes together. The grounding electrode conductor is connected to the water pipe on the interior of the building.

The connection of the grounding electrode conductor to the interior water pipe shall be made a maximum of ___ feet from the point where the water pipe enters the building.

(a) 3 (b) 5 (c) 8 (d) 10

LIST CODE SECTION ________________.

34. A metal elbow installed underground in a run of nonmetallic conduit is not required to be grounded, if it is isolated by a minimum cover of at least ___ inches to any part of the elbow.

(a) 6 (b) 12 (c) 18 (d) 24

LIST CODE SECTION ________________.

35. In indoor wet locations where rigid nonmetallic conduit is installed on a concrete or tile wall, the conduit must have at least a ___ inch airspace between it and the wall.

(a) 0 (b) 1/8 (c) 1/4 (d) 1/2

LIST CODE SECTION ________________.

36. Preassembled cable in nonmetallic conduit is not permitted ___.

(a) in cinder fill
(b) in exposed indoor location
(c) encased or imbedded in concrete
(d) in underground locations subject to severe corrosive influences

LIST CODE SECTION ___________________.

37. The maximum conductor fill allowed for a nonmetallic wireway shall be ___.

(a) 30 current carrying conductors
(b) 20 percent of the cross sectional area of the wireway
(c) 30 current carrying conductors or 20 percent of the cross sectional area of the wireway, whichever is largest
(d) as specified in Table 1, Chapter 9

LIST CODE SECTION ___________________.

38. Where not listed for other support intervals, nonmetallic wireways shall be supported at maximum intervals of ___ feet.

(a) 3 (b) 5 (c) 8 (d) 10

LIST CODE SECTION ___________________.

39. In a straight run of rigid nonmetallic conduit between securely mounted boxes, expansion joints are required where the computed length change due to thermal expansion or contraction is at least ___ inch or more.

(a) 1/8 (b) 1/4 (c) 3/8 (d) 1/2

LIST CODE SECTION ___________________.

40. Electrical nonmetallic tubing (ENT) is permitted to be installed in a wall or ceiling in a building which exceeds three floors above grade, where the wall or ceiling provides a thermal barrier of material that has a finish rating of at least ___.

(a) 15 minutes (b) 30 minutes (c) 1 hour (d) 1 1/2 hours

LIST CODE SECTION ___________________.

F I N A L E X A M

41. A receptacle which is secured solely by a single screw, installed in a raised cover on a four square box ___.

(a) is prohibited in all cases
(b) is allowed without exception
(c) is allowed only for a receptacle listed for such use
(d) is allowed only when the raised cover is installed on a nonmetallic box

LIST CODE SECTION ________________.

42. A metal pole for supporting a lighting fixture is not required to have a handhole at the base to provide access to wires, if the pole has a hinged base and is a maximum of ___ feet in length.

(a) 8 (b) 15 (c) 20 (d) 25

LIST CODE SECTION ________________.

43. Track lighting which operates at 30 volts or higher shall be installed at least ___ feet above the finished floor.

(a) 3 (b) 5 (c) 8 (d) 10

LIST CODE SECTION ________________.

44. Secondary overcurrent protection of transformers over 600 volts ___.

(a) shall be by an individual overcurrent device
(b) is never required, where protected by an appropriate overcurrent device on the primary side
(c) is permitted to consist of not more than six circuit breakers or sets of fuses grouped together
(d) is always required, even where conditions of maintenance and supervision assure that only qualified persons will monitor and service the transformer installation

LIST CODE SECTION ________________.

45. Circuits supplying fixed outdoor electric deicing equipment are required to have ___.

(a) a ground fault circuit interrupter
(b) ground fault protection of equipment
(c) a disconnect in sight from the equipment
(d) a temperature control device with an "off" position with no positive lockout

LIST CODE SECTION ___________________.

46. Given: A fixed electric space heater without a motor is installed in a multifamily dwelling. The heater has no supplementary overcurrent protection.
The heater is controlled with a thermostat which does not have a marked "off" position.
The branch circuit switch or circuit breaker is not "within sight from" the heater.
For the branch circuit switch or circuit breaker to be permitted to serve as the disconnecting means for the heater, the switch or breaker must ___.

(a) be readily accessible
(b) not control lamps or other appliances
(c) be capable of being locked in the open position
(d) be located within the dwelling unit or on the same floor as the heater

LIST CODE SECTION ___________________.

47. Where GFCI protection is located in the power supply cord for an outdoor portable sign, the ground fault circuit interrupter shall be located within ___ inches of the attachment plug.

(a) 6 (b) 12 (c) 18 (d) 24

LIST CODE SECTION ___________________.

48. In a Class I, Division II location where provision must be made for limited flexibility, such as at motor terminals, which of the following wiring methods is not allowed?

(a) flexible metal conduit
(b) armored cable with approved fittings
(c) liquidtight flexible nonmetallic conduit
(d) flexible cord approved for extra hard usage

LIST CODE SECTION ___________________.

F I N A L E X A M

49. At a gasoline dispensing, self service station which is unattended, the emergency disconnect for a circuit leading to dispensing equipment, must be located more than ___ from the dispensers.

(a) 3' (b) 5' (c) 10' (d) 20'

LIST CODE SECTION ________________.

50. The grounded conductor of a feeder circuit serving a recreational vehicle site shall ___.

(a) have an ampacity the same as the ungrounded conductors
(b) have an ampacity of at least 70 percent of the ampacity of the ungrounded conductors
(c) not be smaller than the required grounding electrode conductor based on the size of ungrounded conductors
(d) be sized to carry the maximum unbalance of the load but not be less than two AWG sizes smaller than the ungrounded conductors

LIST CODE SECTION ________________.

IF YOUR NOT ALREADY A MEMBER OF THE IAEI I WOULD STRONGLY SUGGEST APPLYING FOR MEMBERSHIP. THE IAEI HAS ONE OF THE BEST MAGAZINES FOR ELECTRICAL EDUCATION. NUMEROUS CODE QUESTION AND ANSWER ARTICLES IN EACH PUBLICATION.

THE FOLLOWING PAGE HAS AN IAEI APPLICATION FORM AND IS PERFORATED FOR EASY REMOVAL. FILL IT OUT AND START A MEMBERSHIP IN THE INTERNATIONAL ASSOCIATION OF ELECTRICAL INSPECTORS.

I JOINED THE IAEI IN 1984 WHILE I WAS AN ELECTRICAL INSPECTOR FOR WALT DISNEY WORLD.

INTERNATIONAL ASSOCIATION · ELECTRICAL INSPECTORS · Incorporated 1928 · Keystone of the Electrical Industry

International Association of Electrical Inspectors' Major Objectives

- Formulation of standards for safe installation and use of electrical materials, devices and appliances.
- Promotion of uniform understanding and application of the National Electrical Code®, other codes, and any adopted electrical codes in other countries.
- Promotion of uniform administrative ordinances and inspection methods.
- Collection and dissemination of information relative to the safe use of electricity.
- Representation of electrical inspectors in the electrical industry, nationally and internationally.
- Cooperation with national and international organizations in the further development of the electrical industry.
- Promotion of cooperation among inspectors, inspection departments of city, county and state at national and international levels, the electrical industry and the public.

Contact IAEI membership secretary for information on our special membership categories - Section, National and International Member; Sustaining Member (Bronze, Silver, Gold or Platinum); and Inspection Agency Member.

MEMBERSHIP APPLICATION

for Inspector Member Associate Member Student Member

Annual Membership Dues $40
Students $30

Services from the IAEI for you and your organization include all this and more:

- Subscription to *IAEI News* - Bimonthly magazine with latest news on Code changes
- Local Chapter or Division meetings - New products, programs, national experts answer questions on local problems
- Section meetings to promote education and cooperation
- Promote advancement of electrical industry
- Save TIME and MONEY by having the latest information

PLEASE PRINT

Name - Last First M.I.	Chapter, where you live or work, if known
Title	(Division, where appropriate)
Employer	For Office Use Only: Section No. Chapter No. Division No.
Address of Applicant	If previous member, give last membership number
City State or Province Zip or Postal Code	and last year of membership 19____
Student applicants give school attending **	Graduation Date __/__/__
Endorsed by	Endorser's Membership Number
Applicant's Signature	(Area Code) Telephone Number

☐ Check ☐ Money Order ☐ Visa ☐ MasterCard ☐ Amex ☐ Discover Name on Card Charge Card Number Exp. Date

☐ Inspector $40 ☐ Chief Electrical Inspector * $40 ☐ Associate $40 ☐ Student ** $30

Inspector Members and Chief Electrical Inspector Members MUST sign below:

☐ I, ______________________________, meet the qualifications for inspector member type as described below.

Inspector members must regularly make electrical inspections for preventing injury to persons or damage to property on behalf of a governmental agency, insurance agency, rating bureau, recognized testing laboratory or electric light and power company.

*Chief electrical inspector members must regularly provide technical supervision of electrical inspectors.

**Student member must be currently enrolled in an approved college, university, vocational technical school or trade school specializing in electrical training or approved electrical apprenticeship school.

Mail to: IAEI, P. O. Box 830848 Richardson, TX 75083-0848 For information call: (972) 235-1455 (8-5 CST)

ANSWERS

CHAPTER 2 - EXAM #1 Drilling and Notching

1. **(a)** 1/3
2. **(b)** 25%
3. **(a)** 40%
4. **(a)** 1 1/8" & 2 3/8"
5. **(c)** 1/6
6. OKAY
7. VIOLATION
8. VIOLATION
9. VIOLATION
10. OKAY

CHAPTER 2 - EXAM #2 Boxes and Raceways

1. **(b)** separate box 370-28d
2. **(c)** 8 times 370-28a1
3. **(c)** 1 1/4" Tables 4 & 5
4. **(b)** 2 cubic inch T.370-16b
5. **(c)** cover 370-25
6. **(b)** 1/8" 370-21
7. **(d)** I,II,III, & IV 370-23c
8. **(b)** twice 370-16c
9. **(c)** 1/4" 370-17c
10. **(a)** round 370-2
11. **(c)** 15/16" 370-24
12. **(d)** not at all 370-20
13. **(c)** 10 cubic inch T.370-16a
14. **(b)** #4 370-28a
15. **(d)** 3 conductors T.370-16a
16. **(c)** 1/4" 370-18

CHAPTER 3 - EXAM #1

1. **VIOLATION** 230-30 ex.d Bare aluminum must be part of a cable assembly
2. **OKAY** 230-7 ex.1
3. **OKAY** 230-2(d)
4. **VIOLATION** 230-24a
5. **VIOLATION** 230-54c ex.
6. **OKAY** 230-43
7. **OKAY** 230-30 ex.
8. **FALSE** 230-79a
9. **TRUE** 230-93
10. **FALSE** 230-71b
11. **FALSE** 230-9
12. **FALSE** 230-23c Table 250-66
13. **FALSE** 230-24a ex.3
14. **TRUE** 230-90b
15. **TRUE** 230-95c

CHAPTER 4 - EXAM #1 Grounding

1. **true**
2. **true**
3. **(b) II only**
4. **false**
5. **(d) I, II, III**
6. **false**
7. **(d) I, II, III**
8. **false**
9. **(d) 25 Ω**
10. **(b) II only**
11. **(c) 5/8"**
12. **(b) 8 feet**
13. **true**
14. **false**
15. **false**

CHAPTER 4 - EXAM #2 Grounding

1. **(a)** 250-24b
2. **(c)** 250-106 FPN #2
3. **(a)** 250-116 FPN
4. **(b)** 250-102c
5. **(d)** 250-64c
6. **(b)** 250-64b
7. **(c)** 250-122b
8. **(d)** 250-50d
9. **(a)** 250-50c
10. **(c)** 250-50c
11. **(b)** 250-66b
12. **(c)** 250-126(1)

CHAPTER 4 - EXAM #3 Grounding

13. 250-64e **VIOLATION** must be bonded on both ends of the conduit protecting the conductor.

14. Table 250-66 **OKAY** #1/0 phase conductor = **#6 copper** bonding jumper.

15. 250-54 **VIOLATION** must have an equipment grounding conductor.

16. 250-64b **VIOLATION** cannot use flex for protection.

17. 250-148a **VIOLATION** the grounding conductor shall be connected to a screw which serves no other purpose than to maintain continuity of the grounding path. This screw is also used to secure the cables with the clamp.

18. 250-60 FPN2 250-92 **VIOLATION** all system electrodes are to be bonded together. This is an area often violated in the field.

19. 250-94 **OKAY** EMT made up tight is permitted on services.

20. 250-148a **VIOLATION** the connection of a grounding conductor and a metal box shall be to a screw that serves no other purpose than to maintain the continuity of the grounding path to the metal enclosure. Violated everyday in the field!

21. 250-146a **VIOLATION** unless listed to maintain continuity; it is a violation.

22. Table 250-66 **#4 copper** 250-66a **#6 copper from water pipe to ground rod**

23. 250-66 **#4 copper**

24. Table 250-122 **#2 copper** The #1/0 minimum from 310-4 for paralleling does not apply to equipment grounding conductors per 310-4.

25. 3 x 500 = 1500 kcmil x 12 1/2% = 187, 500 cm Tablc 8 rcquircs a **#4/0 main bonding jumper** @ 211, 600 cm. Equipment bonding jumper is the same size **#4/0**.

26. Table 250-66 **#1/0 copper**

27. #400 THW aluminum from Table 310-16 = 270 ampacity. This requires a #300 THW **copper** with equivalent ampacity, a #300 THW copper has an ampacity of 285. Table 250-66 for #300 kcmil copper = **#2 copper** bonding jumper.

28. 600 kcmil x 2 = 1200 kcmil. Table 250-66 over 1100 kcmil = **#3/0 copper** to the water pipe. 250-94 ex.1a **#6 copper** to the ground rod.

29. Table 250-122 1000a CB = **#2/0 copper** equipment grounding conductor in each PVC conduit for the feeder.

30. 250-122 When conductors are adjusted in size to compensate for voltage drop, equipment grounding conductors shall be adjusted proportionately to circular mil area.
750/600 = 1.25 increase in cma
Table 250-122 shows a #3 copper required for a 400 amp circuit breaker.
Table 8 #3 = 52,620 cm x 1.25 increase = 65,775 cm required. A **#2** at 66,360 cm is required.

CHAPTER 5 - EXAM #1 Plan Reading

1. (b) two
2. (c) fourteen
3. (d) A-11

CHAPTER 5 - EXAM #2 Plan Reading

1. (b) two
2. (c) fifteen

CHAPTER 5 - EXAM #3 Plan Reading

1. (b) eight
2. (b) two
3. (a) zero
4. (d) twenty-six
5. (a) seven
6. (b) four
7. (d) six

CHAPTER 5 - EXAM #4 Plan Reading

1. (b) thermostat
2. (a) one
3. (d) five
4. (a) four
5. (b) three
6. (c) seven
7. (a) four

CHAPTER 6 - EXAM #1 Fault Currents

1. (b) 2450 amperes
2. (a) one-half cycle
3. (b) 32,800 amperes
4. (b) 3.5%
5. (d) 200,000 amperes
6. (a) 10,000 amperes

CHAPTER 7 - EXAM #1 Pools

1. (a) 5' 680-70
2. (d) I,II, & III 680-22a2,5 & 23a
3. (d) #12 680-25b1
4. (a) 5' 680-6a1
5. (d) I,II & III 680-24
6. (a) one more 680-21d
7. (b) 4' 680-21a5
8. (d) 10' 680-51e
9. (b) 10' 680-8
10. (c) 18" 680-20a3
11. (b) 5' 680-22a5
12. (c) grd. terminal service equip. 680-25d
13. (b) 15v 680-20a1
14. (d) 5' 680-12
15. (a) #8 680-22b

CHAPTER 8 - EXAM #1 Emergency Systems

1. (c) two hours fuel 700-12b2
2. (d) I,II, & III 702-2 FPN
3. (c) accessible only 700-25
4. (b) 10 seconds 700-12
5. (d) 87 1/2% 701-11a
6. (b) II only 700-26
7. (b) 1 1/2 hours 700-12a
8. (d) I,II, & III 700-7a,b,c
9. (d) I,II, & III 700-12b1,2,3
10. (c) written record 700-4d
11. (c) I & II only 700-16
12. (d) one minute 701-11

CHAPTER 9 - EXAM #1 SIGNS

1. (b) 30a 600-5b2
2. (a) 1/4" 600-41c
3. ([illegible]) 2" 600-9c
4. (d) I,II, & III 600-9d1,2,3
5. (d) 15' 600-10d2
6. (a) outlet 600-6a
7. (d) 20' 600-32j
8. (b) 20a 600-5b1
9. (b) within sight 600-6a
10. (a) listed 600-3
11. (c) GFCI 600-10c2
12. (c) electric discharge 600-4(a)

CHAPTER 9 - EXAM #2 HAZARDOUS

1. (a) general-use type 501-6b2
2. (a) 18" 501-5a1
3. (a) gases & vapors 500-7
4. (a) rigid metal 501-4a
5. (b) ex. moisture 501-5
6. (b) Class II, Div. I 500-8a
7. (b) 24" 514-8 ex.2
8. (d) not allowed 502-2a3
9. (d) liquid-tight 502-4a2
10. (c) Class III, Div. II 500-9b
11. (b) Class I, Div. II 511-3a
12. (c) gasoline pump 514-5a

CHAPTER 9 - EXAM #3 HAZARDOUS

1. (d) water cooled 503-6
2. (a) comb. dust 500-5b
3. (d) I,II, & III 511-6a
4. (d) not classified 511-2
5. (d) EMT 502-4
6. (b) either side 501-5b2
7. (b) intrinsically safe 504-2 DEF
8. (d) not classified 513-3d
9. (c) flexible metal 501-4b
10. (d) 2" 501-5a1
11. (c) 12' 511-7b
12. (d) first fitting 514-7a

CHAPTER 10 - EXAM #1 MISCELLANEOUS INSPECTIONS

1. (b) #20 620-12a1 ex.
2. (a) current-carrying 520-27b
3. (c) 3" 810-54
4. (b) 8.19a 630-31a2 21a x .39
5. (d) 100a 550-23c
6. (b) MI 518-4
7. (b) protector 800-30b FPN
8. (b) 620-91c
9. (c) #6 800-40d
10. (d) I,II, & III 551-2 FPN
11. (a) dressing rooms 520-73
12. (d) 6' 820-10f3

CHAPTER 10 - EXAM #2 MISCELLANEOUS INSPECTIONS

1. (c) I & III only 610-21f4
2. (d) 30a 530-14
3. (a) 1/2" 520-43a
4. (b) 400% 530-18a
5. (b) 300% 630-32a
6. (b) II only 550-5a
7. (a) trolley frame 610-61
8. (d) I,II, & III 530-41
9. (c) #8 540-13
10. (d) #14 640-7a
11. (d) 15' 525-12b
12. (b) II only 670-2

CHAPTER 12 - FINAL EXAM 50 QUESTIONS

1. **(d)** 90-2 (b) 5
2. **(c)** article 100 def.
3. **(d)** article 100 def.
4. **(d)** 110-26 (d)
5. **(d)** 110-31 (c) ex. & 370-72 (e)
6. **(b)** 110-31
7. **(c)** 210-7 (d) 3 (b)
8. **(a)** 210-8
9. **(c)** 210-52a3
10. **(a)** 210-52 (c) 5 ex.
11. **(a)** 210-52 (b) 2
12. **(b)** 250-134 (b)
13. **(b)** 250-97 ex (b)
14. **(c)** 250-104(a)4
15. **(d)** 305-4 (b)
16. **(b)** 305-6 (a)
17. **(b)** 318-6 (j)
18. **(b)** 370-23 (f) ex. 2 (e)
19. **(a)** 374-9 (e) 2 (b) 2
20. **(c)** 410-101b
21. **(b)** 410-57 (b) 2
22. **(b)** 411-6
23. **(a)** 525-30
24. **(a)** 625-22
25. **(b)** 680-4 def.
26. **(c)** 110-26 (e)
27. **(a)** 110-14 (c)
28. **(a)** 210-3
29. **(b)** 210-25
30. **(d)** 210-52 & 424-9
31. **(c)** 210-63
32. **(a)** 210-8 (a) 2 ex.
33. **(b)** 250-50
34. **(c)** 250-80 ex.
35. **(a)** 300-6 (c)
36. **(b)** 343-4 (1) (2)
37. **(b)** 362-19
38. **(a)** 362-22
39. **(b)** 347-9
40. **(a)** 331-3 (2)
41. **(c)** 410-56 (F) 3
42. **(c)** 410-15 (b) 1 ex. #2
43. **(b)** 410-101 (c)
44. **(c)** T. 450-3 note 2
45. **(b)** 426-28
46. **(c)** 424-19 (b) 1
47. **(b)** 600-10 (c) 2
48. **(b)** 501-4 (b)
49. **(d)** 514-5 (c)
50. **(a)** 551-73 (d)